Handbook
of
Experimental
Stomatology

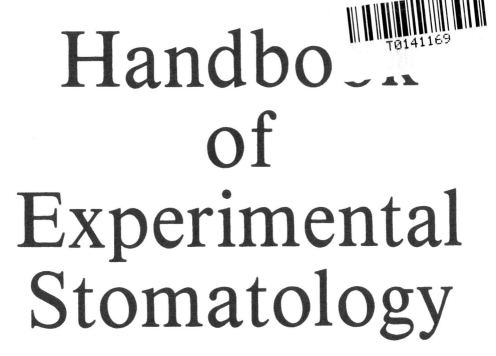

Editors

Samuel Dreizen, D.D.S., M.D.
Professor and Chairman
Department of Dental Oncology
Professor
Department of Pathology
University of Texas Dental Branch
Professor
University of Texas Dental Science Institute
Houston, Texas

Barnet M. Levy, D.D.S., M.S.
Professor and Director
University of Texas Dental Science Institute
Professor of Pathology
University of Texas Dental Branch and
Graduate School of Biomedical Sciences
Houston, Texas

CRC Series in Experimental Oral Biology

Editor-in-Chief
Barnet M. Levy

CRC Press
Taylor & Francis Group
Boca Raton London New York

CRC Press is an imprint of the
Taylor & Francis Group, an **informa** business

CRC Press
Taylor & Francis Group
6000 Broken Sound Parkway NW, Suite 300
Boca Raton, FL 33487-2742

Reissued 2019 by CRC Press

A Library of Congress record exists under LC control number:

Publisher's Note
The publisher has gone to great lengths to ensure the quality of this reprint but points out that some imperfections in the original copies may be apparent.

Disclaimer
The publisher has made every effort to trace copyright holders and welcomes correspondence from those they have been unable to contact.

ISBN 13: 978-0-367-24325-8 (hbk)
ISBN 13: 978-0-367-24327-2 (pbk)
ISBN 13: 978-0-429-28175-4 (ebk)

Visit the Taylor & Francis Web site at http://www.taylorandfrancis.com and the CRC Press Web site at http://www.crcpress.com

THE EDITORS

Dr. Samuel Dreizen is Professor and Chairman, Department of Dental Oncology; Professor, Department of Pathology, University of Texas Dental Branch; Professor, University of Texas Dental Science Institute and Faculty Member, University of Texas Graduate School of Biomedical Sciences at Houston. A native of New York City, Dr. Dreizen is a graduate of Brooklyn College (B.A.), Western Reserve University (D.D.S.) and Northwestern University (M.D.) with academic honors from each institution. For 15 years he was Assistant Scientific Director of the Nutrition Clinic, Hillman Hospital, Birmingham, Alabama. From 1947 through 1966 he was Instructor to Associate Professor, Department of Nutrition and Metabolism, Northwestern University Medical School and from 1948 to 1953 a Clayton Fellow in Nutrition Research. Dr. Dreizen is Consultant in Nutrition to the Council of Dental Therapeutics of the American Dental Association and on the Editorial Boards of *Postgraduate Medicine, Journal of Dental Research*, and *Pharmacology and Therapeutics in Dentistry*. He holds the Edgar Martin Memorial Award of the Odontographic Society of Chicago and has published more than 160 papers in the fields of oncology, stomatology, nutrition, and human growth and development. Dr. Dreizen is the author of *The Mouth in Medicine* published by McGraw-Hill, Inc. in 1971 and 1978.

Dr. Barnet M. Levy is Professor of Pathology, University of Texas Dental Branch at Houston; Adjunct Professor of Veterinary Anatomy, College of Veterinary Medicine, Texas A&M University at College Station; and Adjunct Foundation Scientist, Southwest Foundation for Research and Education at San Antonio, Texas. He is the Director of the University of Texas Dental Science Institute at Houston. He has A.B. and D.D.S. degrees from the University of Pennsylvania and an M.S. from the Medical College of Virginia. He has been President of the International Association for Dental Research, the American Academy of Oral Pathology and the American Board of Oral Pathology. He is Editor of the *Journal of Dental Research*. Dr. Levy has been the recipient of the I.A.D.R. Isaac Schour Memorial Award for Research, the Fred Birnberg Research Medal of Columbia University, and the Alumni Award of Merit from Washington University and the University of Pennsylvania. He is or has been a consultant to many government agencies and research hospitals. He has published more than 140 research papers in the scientific literature.

Dedication

To Nathan Dreizen and Henry I. Levy whose memories we cherish and whose comradeship and brotherly love we sorely miss.

TABLE OF CONTENTS

EXPERIMENTAL STOMATOLOGY

INTRODUCTION

Oral biology encompasses an identifiable core of subdisciplines classified as the basic oral sciences. Oral biology thus, constitutes the multidisciplinary base for dental research. The current body of knowledge is derived from the application of biomedical concepts and disciplines to oral and dental problems. There are now definable limits to establish oral biology as a scientific discipline in its own right — a biodental science so to speak.

Stomatology has been defined as "that branch of medicine which treats of the mouth and its diseases." No other part of the body contains such a diversity of anatomic structures concentrated in one area as the mouth. No other site is the focal point for the expression of so many systemic and local diseases as is the oral cavity. No other region serves so strikingly as an external reflection of internal derangement. No other combination of easily accessible soft and hard tissues is so readily amenable to experimental manipulation. Small wonder, then, that the mouth has so often been selected as an investigative arena for the elicitation of biologic principles and pathogenic mechanisms that apply equally to the entire body.

The objective of all biomedical experimentation is to examine critically the validity of hypotheses in a suitable model under precisely controlled and reproducible conditions. Experimental design must be consistent with the objectives. Methodology must be qualitatively specific and quantitatively sensitive to assure a fair and honest assessment of the test premise. Plausible and conservative interpretation of data derived from appropriate experimental trials is the very keystone of the scientific method.

Experimental stomatology has made substantial contributions to the knowledge explosion that has been the hallmark of the post-World War II era. It has provided a means of reducing pathological events to examinable components. It has proved a tempered tool for distinguishing between fact and fancy and a versatile vehicle for passage from the realm of speculation into the realm of reality. It has become the critical connection between the laboratory and the clinic in the diagnosis, treatment, and control of oral disease.

The present volume is not intended as a complete compendium on the subject of experimental stomatology. The contents have been deliberately confined to experimental oral nutrition, experimental oral medicine, and experimental oral oncology. Future volumes in this series will be devoted to many of the other facets of oral biology. In each, the theoretical will be integrated with the practical by means of the experimental, to provide a concise overview of dental science.

EXPERIMENTAL ORAL NUTRITION

Although no part of the body is immune to inanition, overt reactions to nutritional deprivation are usually site specific. In the clinically apparent target areas, cellular replication, growth, and maturation are slowed or suspended, cellular function is impaired, and cellular morphology is altered. Nowhere is this more vividly evidenced than in the oral cavity where an extremely rapid rate of cell division and cell replacement buttressed by repeated exposure to mechanical, chemical, microbial, and thermal insults, renders the mouth structures particularly prone to disruption by nutritional deficits. The broad expanse of oral mucosa acts as a window for the direct visualization of ongoing nutritionally induced disturbances in tissue integrity. The teeth provide a permanent historical record of those nutritional defects that interfered with mineralization during the formative period. The mouth thus acts as a mirror of past and present nutritional derangement.

The manifestations of malnutrition in the oral structures are nonpathognomonic. They cannot be differentiated from lesions produced by nonnutritional etiologies merely on the basis of gross appearance. In many instances, the lesions may be the result of nutritional and nonnutritional factors operating in the same mouth at the same time. Naturally occurring human nutritional deficiencies are rarely, if ever, absolute or singular. Consequently, details of the manner in which specific nutritional deficits affect the oral structures on the cellular level have come from animal and human experiments in which a single nutrient was the only variable. Elucidation and identification of the 40 to 50 food chemicals that constitute the essential nutrients for man that participate in vital metabolic processes led to the development of chemically defined diets for experimental studies of the macroscopic and microscopic manifestations of single nutrient lack in the oral tissues. Experimentally induced single nutritional deficiencies have also been produced in man and in animals by administration of antimetabolites of the specific nutrient. Antimetabolites differ in molecular arrangement from the corresponding metabolite to a degree sufficient to act as a competitive inhibitor. The inhibitory effect is reaction specific and often may be reversed by providing an excess of the displaced metabolite or by supplying the product of the suppressed reaction. These two relatively simple experimental procedures, namely, altering diets of known chemical composition by single or multiple nutrient deletion and adding chemicals that competitively crowd out essential metabolites applied to appropriate experimental models have yielded exceedingly detailed information of the role of various food factors in oral health and disease. For reasons of propriety, much of the experimental data have been derived from animal rather than from human studies.

Most of the specifics of the stomatologic effects of nutritional impoverishment have emerged from studies of experimentally induced nutritional deficiency states. In each, the study design was to subject nutritionally healthy humans or animals to dietary manipulations that produced the desired deficiency. As demonstrated in the following, all were performed under proscribed conditions that achieved the end stage more rapidly, more frequently, and more unadulteratedly than usually possible under natural conditions.

PROTEIN DEFICIENCY

Human Protein Studies

Protein deficiencies in man are invariably accompanied by calorie and/or vitamin deficiencies. To elucidate the speed and sequence of changes in the human tongue during protein and vitamin depletion and repletion, Stein and Spencer[1] placed three patients in a metabolic research ward on a strictly controlled diet of rice and fruit. The diet was adequate in calories but deficient in protein and in fat and water-soluble vitamins. The protein intake was restricted to 26 g/day and the vitamin A, B complex, C, and D content to 75% or less of the minimum daily requirements.

The development of the deficiency changes in the tongue were slow and progressive, spanning a period of several weeks. Minimal objective changes were recognizable in the three patients by 39 days. The major findings were alterations in the normal pattern of the filiform papillae that started at the tip of the tongue and progressed slowly to other areas of the dorsum. The depression in numbers of filiform papillae was gradual, culminating ultimately in complete disappearance. In contrast, the fungiform papillae were less affected by the deficiency state and became more prominent due to the reduction in number of filiform papillae. In the phase of vitamin repletion (thiamine, riboflavin, niacin, pyridoxine, pantothenate, folic acid, vitamin B_{12}, vitamin C, vitamin A, and vitamin D — given daily in supplemental form), papillary recovery was rapid. Objective signs of improvement were apparent within a few days after the vitamin infusions. The time interval needed for recovery following i.m. administration of the same doses of the same vitamins was only slightly longer. All deficiency changes in the tongue were reversed by the vitamin supplementation even in the presence of the low-protein intake. Since none of the vitamins was administered as a single therapeutic agent, the precise cause of the deficiency glossitis was not determinable in this study of combined protein and vitamin depletion.

Squires[2] found that buccal mucosal scrapings from children with marasmus and kwashiorkor showed typical changes consisting of cellular fragmentation, poor staining, and degeneration when compared to scrapings from healthy children. That these changes resulted from protein-calorie malnutrition was proved experimentally in piglets.[3] Nine litter mate pigs, 3 weeks of age, were fed a basal low-protein (6%) diet. Four received the experimental diet alone; two, the basal diet in which 20% casein replaced an equal amount of starch; three, the basal diet supplemented with 100 g carbohydrate daily for 2 weeks and with 200 g/day thereafter. All were given 300 g of diet daily for the first 2 weeks, 400 g/day from day 15 to 31, and 300 g each subsequent day. Following the onset of deficiency manifestations, one pig in each of the low-protein groups was given an additional 20% casein per day.

Smears were obtained by gentle firm scraping of the buccal mucosa twice weekly during the course of the study. The samples were diluted with a drop of saline, smeared, fixed, and stained. At least 500 cells were counted in each smear. Counts were made in triplicate. All fragments of broken cells containing nuclei were scored. Animals on the low-protein diet plus 20% casein (control group) showed a range of 7 to 14% broken and mutilated cells at each sampling period. Those on the low-protein diet with and without supplemental carbohydrates were initially in the same range. The proportion of mutilated and broken cells increased slowly in those given added carbohydrate and more rapidly and extensively in those with no supplemental carbohydrate. Increasing the protein content of the diet resulted in a prompt reversal of the counts to control values. The findings in the pig mucosa were similar to those reported in African children suffering from marasmus and kwashiorkor.

Animal Protein Studies

An adequate supply of protein is so vital for cellular metabolism that any prolonged protein deficit results in widespread tissue disruption. The effects of protein deprivation on the oral structures of the rat were studied by Stahl et al.[4] In the study, 32 adult Wistar strain white rats weighing 375 g were divided into three dietary groups. Each diet was essentially the same in minerals, vitamins, carbohydrates, and fats but differed widely in protein content. These ranged from 23% protein (Group I) to 2% protein (Group II) to 0% protein (Group III).

Grossly, animals on the protein-deficient diets (Groups II and III) developed atrophy of the lingual papillae. Histologically, the covering epithelium was thinned, and only rare mitoses were present in the germinal layer. The underlying connective tissue appeared edematous. Atrophy of the muscle bundles was found in some sections. The buccal epithelium of the protein-deprived animals evidenced shedding of cells and degeneration of the vascular beds of the corium, denoted by endothelial proliferation and swelling. Many of the muscle bundles were atrophic; some of the muscle fibers had evidence of degeneration. Deficiency manifestations in the periodontium included osteoporosis of the alveolar bone and swelling, disorganization, and pale staining of the periodontal ligament fibers. All of the abnormalities disappeared following protein repletion.

Person et al.[5] examined the histologic response of the oral tissues of rats to protein deprivation. In this study, 30 male Wistar adult rats with an average weight of 320 g were divided into three groups of 10 each. Group I was fed a protein-free diet; Group II was pair-fed an isocaloric 18% casein diet; Group III was given the isocaloric 18% casein diet *ad libitum*. All animals were followed for 62 days after which they were anesthetized, sacrificed, and necropsied. Tissues were fixed in 10% formalin and processed for histologic examination.

The terminal weight of the protein-depleted animals was 46% less than that of the *ad libitum* fed controls. With the exception of the liver, which showed extensive fatty metamorphosis, all organ weights were lower in the protein-poor animals than in the controls. The greatest weight loss was registered by the salivary glands. Both the salivary gland acini and salivary gland ducts were diminished in size. Spaces indicative of edema were present between borders of adjacent acini. The tongue changes were of a mild atrophic nature and were relegated to the covering epithelium, lingual papillae, and lingual glands.

Ziskin et al.[6] described the epithelial changes in the tongues of young rats confined to a protein-deficient diet. Ten immature albino rats, five males and five females, were fed a synthetic diet complete in all respects except for the omission of protein. The animals lost considerable hair and that which remained was arranged in pointed tufts producing a porcupinelike appearance. They became irritable and developed neurological symptoms that affected their gait. Weight loss was progressive, with the final weights equal to those at weaning (35 to 40 g). Terminally, some showed swellings on the face and torso that proved to be amorphous, colorless, gelatinous masses at autopsy. Histologic examination of the tongue disclosed a reduction of epithelial thickness on the dorsum. The cornified papillae were replaced by smooth noncornified epithelium. The entire epithelial covering of the dorsum resembled that normally found on the undersurface of the tongue. Casein added to the diet of two animals with marked clinical symptoms caused a growth of the lingual papillae.

Johnson et al.[7] placed Sprague-Dawley rats on a standard pelleted stock diet and on a powdered diet adequate in protein for 1 week. Half of the latter was then fed a powdered diet low in protein (0.5% lactalbumin). The parotid glands of the animals on the stock diet increased 46% in DNA; 150 to 200% in weight, RNA, and protein; and 285% in amylase during the 10-week study period. In contrast, the glands of rats given the protein-adequate powdered diet had no change in DNA while undergoing increases of 100% in weight and in amylase, 76% in protein, and 23% in RNA. On the low-protein diet, gland weight, DNA, RNA, and protein decreased 20 to 40% and amylase, 72%. There was no parotid gland enlargement in the protein-deficient rats such as occurs in rats whose daily food intake is reduced by 8 to 30% or in acute onset human undernutrition.

The effect of protein deficiency on alveolar bone density was evaluated histometri-cally in the rat by Pack.[8] In this study, 70 male Wistar rats were equally divided into control and experimental groups fed powdered diets containing 21.7% and 7.6% protein, respectively. Five control and five experimental animals were killed at 4-week intervals for 4 months and at 6, 9, and 12 study months. Examination of demineralized sections of the jaws with a modified Chalkley point graticule allowed histometric assessment of the density of the alveolar trabecular bone in the upper and lower molar areas. Protein deficiency resulted in a significant reduction in bone density manifested as osteoporosis, particularly in the lower jaw. A moderately high correlation between trabecular density and body weight was found in both groups.

The influence of protein deprivation on the periodontal structures of the rat was investigated by Chawla and Glickman.[9] In this study, 80 male and female Wistar rats, 25 days of age and weighing 45 to 60 g, were split into two equal groups fed synthetic diets containing 2% protein (experimental group) and 22% protein (control group).

The diets were isocaloric. Each rat was given 15 g of ration and water *ad libitum* daily. Food consumption was equal for about 4 weeks. During the next 4 weeks, food intake by the protein-depleted group diminished perceptibly. Equal numbers of control and experimental animals were killed at 2-week intervals throughout the 8-week study period. At necropsy, the jaws, vertebrae, and femurs were removed, fixed in 10% formalin, decalcified, paraffin-embedded, sectioned, and stained with eosin and hematoxylin and with Mallory's trichrome.

The protein deficiency produced degeneration of the connective tissue in the gingiva and periodontal ligaments, osteoporosis of the alveolar bone, and retarded deposition of cementum. Changes in the periodontal ligament and alveolar bone were identical to those in the periosteum of the other bones examined. Protein-deficiency osteoporosis resulted from reduced deposition of osteoid and retardation in the morphodifferentiation of connective tissue cells into osteoblasts.

Goldman[10] compared the effects of protein-free diets on the periodontal tissues of rats and spider monkeys. In the study, 61 rats and 16 spider monkeys were used in the experiments. The rats were assigned to two groups on the basis of age: 26 young animals weighing 150 to 200 g and 35 adults weighing 300 to 350 g. The monkeys were divided into eight juveniles and eight adults. The protein-free diets contained starch, nonnutritive fiber, hydrogenated vegetable oil, salts, cod liver oil, vitamin B complex, and ascorbic acid. The rats were sacrificed at varying times when moribund (70 to 80 days) and the monkeys when in a similar state (115 to 145 days).

Eosin and hematoxylin, Mallory's and Wilder's silver-stained sections of the periodontium showed that protein deprivation prompted degenerative changes in the attachment apparatus of the rat and monkey teeth in both young and old animals. Protein deficiency caused a disappearance of fibroblasts and connective tissue fibers from the periodontal ligament, resulting in a loose edematous appearance and osteoporosis of the supporting bone. Young animals were more severely affected than old presumably because of the higher tissue synthetic activity.

Shaw[11,12] found that offspring of rat dams with marginal protein deficiency throughout pregnancy and lactation had low weaning weights, reduced size of molars, and delayed eruption of and abnormal cuspal patterns in the third molars. Complete protein-deficient diets fed for 5-day intervals in the lactation period also caused reduced weaning weight and delayed eruption of third molars. These influences were greatest when the protein deficiency was late in lactation. Marginal maternal protein deficiency (6% casein) during the second reproductive cycle resulted in offspring with smaller skull dimensions at each age than those in siblings from the first litter when the mothers were fed adequate protein (24% casein). Feeding diets with adequate protein to the offspring of protein-depleted mothers improved the growth of these animals without a catch-up in skull dimensions. The skull dimensions of young adults constituted more permanent evidence than body weight of the marginal protein deficiency imposed on the mothers during pregnancy and lactation. The greatest penalty in body weight and skull dimensions occurred when the marginal protein deficiency was maintained both throughout the reproductive cycle and the rapid growth period in the first few months of life.

Stahl[13] delineated the effects of a low-protein diet on oral wound healing in rats. In this study, 64 Sprague-Dawley female rats averaging 171 g in weight were assigned to three dietary regimens. Group I was given an 8% casein diet *ad libitum*; Group II was pair-fed an isocaloric 27% casein diet with Group I; Group III was offered the 27% casein diet *ad libitum* and served as controls. A like number of adult females averaging 353 g was similarly grouped and fed.

Simultaneous with the initiation of the dietary protocols, a wound was made in each animal by stripping the gingiva distal to the upper left first molar. Representatives from each group were sacrificed immediately after injury and at intervals ranging from 1 to 31 days after wounding. The rats were decapitated and left maxilla dissected out and processed for histologic examination. All of the animals showed epithelialization of the gingival wound in 5 to 8 days. Crestal osteogenesis in the low-protein-fed rats lagged behind the controls throughout the experimental period. Extent of lag was similar in both age groups. Inflammation persisted in some of the protein-deficient animals into the latter stages of wound healing and was often associated with sequestra formation. Inflammation delayed connective tissue organization of the wound, but did not cause disruption of the epithelium.

To determine the influence of a low-protein diet on the wound site after initial healing had begun, Stahl[14] fed 83 male Sprague-Dawley rats, mean weight 255 g, a stock laboratory chow and water *ad libitum* for 1 month after creating a standard wound at the mesial gingival margin of the maxillary left first molar. In the study, 21 animals were then continued on the stock diet; 22 were given an 8% casein diet *ad libitum*; 21 were pair-fed a 27% casein diet isocaloric with the 6% casein diet; 19 were given the 27% casein diet unrestrictedly. The rats were killed in subgroups at monthly intervals over a 6-month period. At necropsy, the right and left maxilla were removed and prepared for histologic study. Epithelialization of the wounds occurred within a similar time frame in all animals. Clinically, all of the wound sites appeared healed. Histologically, there was inflammation and associated alveolar bone sequestration in a larger number of the nutritionally deprived than adequately fed rats. The pathologic changes lingered for a prolonged time period.

The experimental evidence clearly shows that protein deficiency has a profound effect on the mouth structures. Protein deprivation manifestations range from cellular degeneration through atrophy of the salivary glands, lingual papillae, alveolar bone, periodontium, and developing teeth, to delayed oral wound healing. While there is a substantial body of experimental data on the role of protein in maintaining the integrity of the oral tissues, as detailed below, studies of single amino acid deficiencies have been confined almost exclusively to tryptophan and lysine.

AMINO ACID DEFICIENCY

Tryptophan Studies

Bavetta et al.[15] explored the effect of tryptophan deficiency on the jaws and femurs of growing rats. In this study, 40 female weanling rats (21 days old) from 11 litters were randomized either to tryptophan-deficient diet for 3 weeks and repletion for 4 weeks (Group I); tryptophan-deficient diet for 7 weeks (Group II); control diet for 7 weeks (Group III). Each group received a standard diet for 1 week after weaning before starting the experiment. The test diet consisted of an acid hydrolysate of casein (a preparation almost completely free of tryptophan), Mazola® oil, wheat germ oil, salts, starch, and vitamins. The casein hydrolysate in the control diet was supplemented with 0.75% tryptophan. Equal numbers of animals from each group were sacrificed at the end of 3 and 7 weeks. The heads and femurs were removed, fixed, decalcified, and processed for histologic examination.

Both endochondral and periosteal bone formation were inhibited by the deficiency of tryptophan. The alveolar process was affected to a greater degree than the long bones. The mandibular condyle of the deficient animals was smaller in size and had arrested chondrogenesis and osteogenesis, leading to a flattening of the condylar head. Both the radicular and interseptal areas of the alveolar bone were markedly affected. The resulting osteoporosis was characterized by an almost complete disappearance of spongy bone trabeculae. There were localized areas of fragmentation and loss of periodontal attachment fibers in the interradicular bone. Inhibition and retardation of endochondral and periosteal bone formation were the direct result of reduced protein synthesis. This was evidenced by a reduction in width of the calcifying cartilage. All pathologic changes underwent reversal upon repletion with tryptophan.

To establish whether the aforementioned changes would be intensified by prolonged tryptophan deficiency, Bavetta and Bernick[16] extended the experimental period to 26 weeks. By 3 weeks, there was a slight narrowing of the alveolar bone proper and spongy trabeculae and an enlargement of the marrow space in the interradicular bone that became accentuated with time. At 8 weeks, only a shell of alveolar bone remained, consisting merely of the cribriform plate. At 16 weeks, the alveolar bone proper was fragmented into small islands that appeared as isolated osteogenic centers. There was a loss of attachment of many of the periodontal fibers in the interradicular bone and a marked lowering of the crest of the interseptal bone to about one half of the molar root length.

The effects of tryptophan deficiency during lactation on the bones and teeth of the rat were examined by Bavetta et al.[17] Mother rats were placed on tryptophan-limiting diets on the day of littering and for the first 10 days of lactation. They were then given a stock diet up to weaning on day 21. The tryptophan-limiting diets consisted of (1) protein-free, (2) tryptophan-free, (3) 0.25% tryptophan, (4) 0.75% tryptophan, and (5) 18% casein. Four offspring from each experimental group were killed at 10 and 21 days of age. Heads and tibias were removed and processed for histologic study.

Protein deprivation during the first 10 days of lactation was more stressful than complete tryptophan deficiency. Only two of ten mothers had living young after 10 days on protein-free diet. Offspring of animals on the tryptophan-free, protein-free, and 0.25% tryptophan diets had arrested osteogenesis and decreased bone size. There was a retardation in the development, growth, and eruption of the molar teeth in animals whose mothers were fed tryptophan-free or protein-free diets. Delayed development was characterized by persistence of a well-developed bony capsule over the occlusal surface of the tooth and almost complete retention of Hertwig's sheath surrounding the base of the dental papilla.

Induction of hyperthyroidism by feeding of thyroactive substances has been widely employed as an experimental tool for the production or accentuation of nutritional deficiencies. The influence of dietary thyroid on the periodontium of rats with total and partial tryptophan deficiencies was elucidated by Bavetta et al.[18] In their study, 100 female albino rats 21 to 24 days of age were divided into groups of 10 each and fed diets of known chemical composition. These included (1) tryptophan-free, (2) tryptophan-free plus 0.5% USP desiccated thyroid, (3) diets (1) and (2) fortified with 600 mg, 1200 mg, and 2400 mg dl-tryptophan per kg. Experiments were also conducted with diets identical to rations (1) and (2) except that the acid hydrolysate was replaced by intact casein. After 28 days of ad libitum feeding, surviving rats in each group were killed and the heads and long bones processed for light microscopy.

Animals restricted to the tryptophan-free diet failed to survive the experimental period. Diets containing 600 mg or less of dl-tryptophan per kilogram caused inhibition and retardation of endochondral and periosteal bone formation and marked osteoporosis in the alveolar bone around the molar teeth. These effects were intensified by the desiccated thyroid. Ulceration of the gingival epithelium with crater formation developed in the rats receiving dietary thyroid and the tryptophan-free ration supplemented with 600 mg or less of the amino acid. These findings were not observed in animals given identical dietary regimens without thyroid.

Bernick and Bavetta[19] performed histochemical and electron microscope studies of the incisor dentin of rats placed on a tryptophan-deficient diet immediately after weaning and continued for 3 to 26 weeks. Control animals were given the same diet supplemented with 1% *dl*-tryptophan. At sacrifice, the heads were dissected, fixed in Bouin's solution, decalcified, dehydrated, and embedded in nitrocellulose. Sections were stained with toluidin blue for metachromasia, PAS for acid mucopolysaccharides, and Gomori's alkaline phosphatase reaction. Blocks of incisal dentin were prepared for electron microscopy and examined.

The staining reactions indicated that the dentinal ground substance was in a depolymerized state from the absence of tryptophan that prevented normal calcification. EM studies disclosed that the orientation of the collagen fibrils in the deficient dentin was visible, whereas in normal dentin the fibrils were masked by unaltered ground substance. The incisor dentin of the tryptophan-deficient rats contained an increased number of interglobular spaces also indicative of disturbed calcification.

Lysine Studies

Evidence that a deficiency of lysine may detectably affect the dental structures has been provided by Bavetta and Bernick.[20] Young female rats were given a Purina® chow diet for 1 week after weaning and then assigned to one of four experimental groups. Group I received an adequate diet and acted as controls; Group II was fed a lysine-deficient diet for 3 weeks followed by a lysine-rich diet for 4 weeks; Group III was given the lysine-poor diet for 7 weeks and the lysine-rich diet for 4 weeks; Group IV consumed the lysine-poor diet for 11 weeks. At the end of each study period, the animals were killed, and the jaws were excised, fixed, and prepared for staining.

The lysine-deficient rats showed a marked narrowing of the cartilage plate in the mandibular condyle. The transition layer was reduced to two to three cells instead of the normal five to six cells. There was a notable absence of trabeculae in the subchondral region. Chondrogenesis and osteogenesis were resumed when lysine was added to the diet. The interradicular and septal portions of the alveolar bone showed a thinning of the spongiosa at 3 weeks and progressive degenerative changes at 7 weeks. Cribriform plates were disintegrated, the alveolar bone proper was osteoporotic, and there was a breakdown of bone with complete loss of attachment fibers in the interradicular region. During the repletion period, the osteoporosis disappeared and reparative osteogenesis was evident. Incisal dentin deposited during the lysine deficiency was hypocalcified, suggestive of a disturbance in dentin deposition.

A lack of either tryptophan or lysine markedly suppresses the growth and maturation of the osseous, cartilaginous, collagenous, and dentinal connective tissue elements in the jaws of laboratory rodents. Both are part of the indispensable core of ten essential amino acids required for the satisfactory growth of a wide variety of animal species. The manifestations of each essential amino acid deficiency emanate from an interruption in protein biosynthesis, accounting both for the similarity of their characteristic features and for the indistinguishability of these characteristics from protein deficiency per se.

MINERAL DEFICIENCY

Although virtually all presently known chemical elements have been found in living cells, only about 13 different minerals have been shown to be needed by man. All must be provided by the daily diet, some in trace amounts. With the exception of calcium, iron, and iodine, naturally occurring human mineral deficiencies are extremely rare. Consequently, most of the specifics of the roles played by the nutritionally essential minerals in maintaining tissue health have been supplied by experimentally induced deficiency states in appropriate animal models.

Calcium Studies

Messer and Herold[21] compared bone loss from the alveolar bone, vertebra, and femur during acute calcium deficiency in female mice. A mild calcium stress was imposed by gestation and lactation; a severe calcium deficiency, by feeding a low-calcium diet during lactation. Nonpregnant female mice were used as controls. Animals were killed either after parturition, at weekly intervals during lactation, or at 3 weeks after lactation. Alveolar bone loss was assessed from alveolar crest height and cross-sectional area of trabecular bone between the first and second molars. Femur bone loss was measured by ash weight and cross-sectional area at midshaft; vertebral bone loss, by ash weight. During gestation and lactation, the total bone loss was about 20%, with that in the vertebral and alveolar bone greater than in the femur. Severe calcium deficiency resulted in a reduction of about 50% in total skeletal mass, with comparable losses in all sites. Alveolar bone changes were typified by a diminution in trabecular bone and maintenance of alveolar crest height. The pattern did not resemble bone loss usually associated with periodontal disease. Response of the alveolar bone to short-term systemic stress was clearly different from that occurring in periodontal disease.

Ferguson and Hartles[22] delineated the effects of diets deficient in calcium and phosphorus with and without vitamin D on the dentin, cementum, and alveolar bone of young rats. In this study, 24 black and white hooded rats, 21 to 24 days of age, were distributed into six groups comprised of equal numbers of males and females. Group I received a diet adequate in calcium, phosphorus, and vitamin D; Group II, a diet low in calcium; Group III, a diet low in vitamin D; Group IV, a diet low in calcium and vitamin D; Group V, a diet low in phosphorus; Group VI, a diet low in phosphorus and vitamin D. The low-calcium diets contained 0.026% calcium; the low-phosphorus diets, 0.06% phosphorus. Animals were killed after a 14-week study period. Molar blocks from one side of the maxilla and mandible were used for the preparation of decalcified sections; blocks from the other side, for making undecalcified sections.

The formation of secondary dentin was severely disturbed by the double deficiency of calcium and vitamin D. There was a reduction in the amount of calcified secondary cementum and a great increase in the quantity of cementoid. Disturbance in secondary dentin production by the diet deficient in phosphorus and vitamin D, while noticeable, was not so severe. The teeth had an irregular deposition of poorly mineralized secondary cementum surrounded by cementoid. A single deficiency of phosphorus caused a greater disruption of cementum formation than did a deficiency of calcium. Animals on the low calcium-normal vitamin D diet had slightly porotic alveolar bone and a great increase in the width of the osteoid seams. Those on the low phosphorus-normal vitamin D diet had some porosis of the alveolar bone and varying widths of osteoid tissue. Rats on the low phosphorus-low vitamin D diet had severe rickets plus large quantities of osteoid tissue in the alveolar bone.

Ericsson and Ekberg[23] tested the preventive and curative properties of calcium and fluoride in dietetically provoked general and alveolar bone osteopenia in the rat. Male albino animals were fed diets of known nutritional composition that were either adequate or deficient in calcium, low in protein, and high in sucrose beginning at 21 to 100 days of age. For varying parts of the experimental period, the rats were given either distilled water or water containing 2000 ppm Ca or 40 ppm F, or both, to drink.

Assessments were made of femoral and alveolar bone density. Femoral bone density parameters included total weight, ash weight, and calcium and phosphorus content. Alveolar bone density measurements were made on the left mandible after processing for histologic sectioning. Six to seven microsections were made in a plane through the apices and axes of the distal root of the first molar and the mesial and distal roots of the second molar. The best sections were stained with Azan and mounted with Canada balsam. Bone density was assessed in the areas between the first and second molars and between the roots of the second molar by projecting the section at 73 times magnification onto squared paper and recording the percentage number of line groups covered by calcified bone.

Bone density in the animals on the calcium-deficient diet was greatly improved by the ingestion of either calcium or calcium and fluoride in the drinking water. There was a tendency to higher mineral density with calcium plus fluoride than with calcium alone. Rats that had been on a high fluoride-adequate diet were much more resistant to bone mineral depletion by the calcium-deficient diet than were controls fed the adequate diet without fluoride. The alveolar bone reacted much more strongly than the femoral bone to both deprivation and repletion.

Phosphorus Studies

Coleman et al.[24] detailed the manifestations of phosphorus deficiency in the rat skull, teeth, and mandibular condyle. In this study, 49 female Long-Evans rats were arranged in four groups, each with an experimental and pair-fed control arm. The very low phosphorus content of the experimental diet was achieved by using isoelectrically precipitated fibrin for protein and purified phosphorus-free components for the remainder of the diet. The animals were sacrificed in groups with their controls after 24, 32, 41, and 49 study days. At autopsy, the heads were disarticulated, fixed, cut into tooth-bearing blocks, and prepared for histologic examination.

Disturbances in calcification and ossification in the phosphorus deficient rats were very prominent and pronounced in dentin, alveolar bone, and the mandibular condyle. Formation and calcification of incisal and molar dentin were severely retarded. The zone of predentin was abnormally wide, and the pulp chambers were large and contained numerous isolated calcospherites. Molar alveolar bone consisted of large amounts of osteoid deposited on cores of calcified bone. Osteoid formation was apparently unchecked by resorption causing ankylosis of bone and cementum in several areas. The mandibular condyle cartilage failed to calcify. Invasion of connective tissue cells and capillaries were negligible. Cartilage continued to proliferate without being replaced by bone, and the condylar head consisted almost entirely of osteoid tissue. Failure of the formation of a constricted neck region in the condyle was indicative of a lack of remodeling resorption.

Weinmann and Schour[25] determined the effects of phosphate administration on the alveolar bone and dentin of rachitic rats as part of a series of experimental studies in calcification. Single or multiple daily i.p. injections of 2.5 mℓ per 100 g body weight of a mixture of monosodium and disodium phosphate containing 7.5 mg phosphorus per dose were given to 12 animals beginning at age 49 days. They had been on a rachitogenic diet for 28 days. The rats were sacrificed 6 hours to 6 days after the first injection. The jaws were dissected and the upper incisors and molars prepared for histologic sectioning and staining.

The osteoid tissue in the alveolar bone began to calcify at the end of the 2nd day following phosphate dosing. Calcification began initially in the oldest layers. Hyaline tissue formed by degeneration of connective tissue showed the first signs of calcification almost immediately after phosphate treatment. The earliest layers to calcify were those that had formed last. After calcification of the osteoid and hyaline, osteoclasts reappeared and resorbed the calcified portions. Resorption of calcified hyaline progressed at a faster rate than resorption of calcified osteoid. Rachitic dentin showed a resumption of normal calcification and rate of formation following phosphate administration.

Magnesium Studies

Klein et al.[26] investigated the impact of magnesium deficiency on the teeth and supporting structures in the rat. The mouths of 24 rats restricted to a low-magnesium diet at weaning that died or were killed after 3 to 106 days on the experimental regimen were scrutinized clinically and histologically. Magnesium deficiency produced a blanching of the oral mucous membranes. The gingiva was converted into a bulbous mass of smooth whitish-gray tissue that separated the lower incisors. Molar gingiva contained darkly colored brown and yellow irregularly shaped hard material embedded in tissue that engulfed the teeth.

Microscopically, the low-magnesium diet induced a decreased cell count and increased amounts of pink-staining intercellular substances in the periodontium and the formation of a deep blue-staining amorphous material in the periodontal bone after 17 days of feeding. Jaw sections from animals given the diet for 3 months showed loss of bone from around the molar teeth and replacement by large deep-staining masses of tissue containing many spindle-shaped cells. The incisor teeth were surrounded by a large collection of similar tissue.

The disruptions in tooth structure resulting from experimentally instigated magnesium deficiency have been detailed by Becks and Furuta.[27-29] In this study, 14 28-day-old rats were fed a basal diet containing 1.3 to 2.13 mg magnesium per 100 g diet. Control animals received the identical diet supplemented with 0.5% magnesium sulfate. The feeding period extended from 10 to 186 days. Animals were sacrificed at between 50 and 214 days of age. At necropsy, the skull was halved, the mandible disarticulated, fixed, decalcified, celloidin-embedded, sectioned, and stained.

Magnesium deficiency produced distinct and progressive regressive changes in the enamel epithelium, enamel, dentin, and pulp. In the anterior teeth, the ameloblasts showed various stages of localized degeneration and subsequent formation of enamel hypoplasia. The hypoplastic areas increased in number and in size with increasing duration of the deficiency. In some animals, enamel hypoplasia occurred after 20 days of magnesium depletion; all had enamel hypoplasia after 41 days. In the 186-day animal, enamel formation at the apex of the incisor had ceased completely because of the absence of an ameloblastic layer. The underlying connective tissue was in direct contact with the dentinal surface. In the incisal portion, there were deposits of amorphous chromatophilic material in the subepithelial connective tissue. The reduced enamel epithelium usually found in this region was replaced by an acellular structure.

Deficiency-induced dentinal and pulpal changes were more pronounced in the constantly growing anterior teeth than in the molars. In animals on the magnesium-deficient diet for 41 days, only the cortical dentin was well calcified. Pulpally, there were striations indicating a disturbance in calcification. The normal pattern was lost, and the incisal pulp was replaced by an amorphous chromatophilic calcified mass that stained dark and uneven. There were also localized areas of odontoblastic degeneration. After 84 days, the pathologic calcification of the pulp tissue extended apically to include almost two thirds of the tooth. Odontoblastic degeneration was marked. Newly formed dentin showed severe striations and was thrown into numerous folds on the labial side near the apex, indicative of delayed eruption. After 186 days, all of the changes were more severe. There was complete calcification of the entire pulp and cessation of normal dentin formation. The folds at the apex were increased in severity and were now also located on the lingual side.

Gagnon et al.[30] examined the effects of magnesium deficiency on dentin apposition and incisor eruption in the rat. In this study, 17 white rats, 68 to 138 days of age, were placed on a magnesium-deficient diet for 22 to 85 days. Eleven litter mate controls received the same diet fortified with magnesium sulfate.

The earliest deficiency manifestations in the incisors, demonstrable radiographically, was a widening of the periodontal membrane, an indistinct outline of the lamina dura, and a disturbed contour of the enamel surface noted after 22 to 40 study days. These changes became more marked with time as the pulp shifted labially with concomitant thinning of the dentin and folding of the basal portion. The pulp also showed radiopaque longitudinal streaks. In the advanced stages, the molars had a widened periodontal membrane, and there was rarefaction of the alveolar bone, with the roots standing out clearly by contrast. Incisal eruption rate was reduced to approximately one third normal. Dentin apposition was regressively decelerated, with the enamel-covered dentin being affected more acutely than the cementum-covered dentin. There were temporary local cessations of dentinal growth. Formation of the alveolar bone was slowed to one half or less of the normal rate.

Irving[31] not only determined the dental changes during magnesium deficiency but also studied the recovery process following addition of the element to the diet. In the study, 39 rats were fed a magnesium-deficient regimen. Four or more were killed and examined after 2, 4, and 6 days. Some of the remaining rats were transferred to an adequate diet, killed, and assessed after 2 to 10 days, while others were kept on the low-magnesium diet until death. Rats fed the adequate diet from weaning constituted the control group. Longitudinal sections were made of the upper incisors and stained with hematoxylin and eosin.

The first characteristic change of magnesium deficiency, an increase in the width of the predentin, occurred in 4 days. Predentin width increased to about three times normal as the deficiency continued. Dentin formed during this period had a translucent appearance, stained lightly with hematoxylin, and contained unduly prominent tubules. After 8 days, the dentin showed definite striations. The odontoblasts underwent progressive atrophy and were embedded in the predentin. Large calcareous granules developed in the enamel organ after 6 days. Subsequently, they increased in frequency and caused marked distortion. After 23 days the entire enamel organ was atrophic. Changes in the molar teeth were relegated to some stratification in the dentin. In animals realimented with dietary magnesium, there was a slow replacement of the abnormalities with normal tissue.

Bernick and Hungerford[32] delineated the effect of magnesium deficiency on various components of bone and dentin by histochemical technics. Male Holtzman rats weighing 120 to 140 g were distributed into two groups. Group I (ten animals) was supplied with a nutritionally adequate diet; Group II (ten animals), with the identical diet devoid of magnesium. The animals were given demineralized water for drinking and were sacrificed after 19 study days. At necropsy, the heads and tibias were removed, fixed, dehydrated, and infiltrated for nitrocellulose embedding. Sections were stained with hematoxylin and eosin, Mallory's connective tissue stain, periodic acid-Schiff reaction, Pearson's silver nitrate impregnation, and toluidine blue.

The epiphyseal cartilage of the posterior head of the tibia was characterized by a widened hypertrophic zone and a narrow proliferating zone. Diaphyseal trabeculae and incisal dentin stained deep red with PAS and exhibited an abundance of Alcian blue positive material that also stained intense violet with toluidine blue. These staining reactions were indicative of an interference with the calcification of cartilage, bone, and dentin. Lack of magnesium presumably produced the morphologic and histochemical changes by inhibiting magnesium-dependent enzyme activity, thereby leading to alterations in collagen, protein, and mucopolysaccharide state in bone and dentin and preventing deposition of bone salts in these structures.

Zinc Studies

Barney et al.[33] produced a zinc-deficiency lingual parakeratosis in the squirrel monkey (*Saimiri sciureus*) by restricting 11 weanling animals weighing 300 g or less to a diet composed of low-zinc casein, cottonseed oil, zinc-free minerals, vitamins, and sucrose. A second group of monkeys given the same diet supplemented with zinc sulphate acted as controls. All were housed in individual plastic cages. Drinking water and food were provided in plastic cups. Animals on the low-zinc diet lost weight while those given supplemental zinc thrived. The monkeys were sacrificed and necropsied after 60 to 330 study days.

All of the monkeys fed the low-zinc diet had pathologic changes in the tongue and palate. Lingual manifestations consisted of a distinct thickening of the mucosa that was particularly prominent over the anterior dorsum and posterior part. Mucosal cells were increased in number. Those in the basal layer were crowded, had hyperchromatic nuclei and formed large rete pegs extending deep into the submucosa, imparting a clubbed appearance in some regions. Parakeratosis was evidenced by a paucity or absence of keratohyaline granules in the cell layers subjacent to the filiform papillae. The underlying muscle fibers were noticeably atrophic. Zinc-deficiency-associated hyperplastic mucosal changes were also present in the posterior part of the roof of the mouth.

Catalanotto and Nandra[34] tested the hypothesis that blockage or coverage of the taste pore regions may result from zinc-deficiency-induced hyperkeratosis of the fungiform and circumvallate papillae. In a previous study using a two-bottle preference test and various concentrations of different tastants, these investigators had found pronounced alterations in the intake of table salt, sucrose, quinine sulphate, and hydrochloric acid in zinc-deficient rats, supporting the proposition that zinc depletion results in altered taste acuity. Accordingly, taste preference, fluid intake, and histology of the tongue epithelium were studied in rats on a zinc-deficient diet and in pair-fed controls.

Results of the two-bottle preference tests showed that zinc-deficient rats had a significantly increased preference for sour and bitter and a significant increase in relative fluid intake. Histologic examination of the tongue epithelium of the zinc-deficient animals revealed a hypercellular prickle cell layer and marked parakeratosis. The fungiform papillae exhibited acanthosis and mild parakeratosis, but the taste bud region was not covered by parakeratotic epithelium sufficient to cause blockage of the tastant from the pore region.

Meyer and Alvares[35] performed a quantitative study of the dry weight and size of cells in the buccal epithelium of zinc-deficient rats. A diet containing 1.3 mg zinc per kilogram was given to male weanling rats for 4 weeks. Controls received the same diet supplemented with 40 mg zinc per kilogram. After death, dry weight of the buccal epithelium was determined for successive histologic layers by ultramicroassay technics and the relative size of the granular cells assayed by a comparison of the number of cell nuclei in the experimental and control groups.

The buccal epithelium of the zinc-deficient animals was parakeratotic and evidenced increases in epithelial thickness and in number of dividing cells. Dry weight per volume of basal and granular cells was higher in the zinc-deprived animals than in the controls. Rate of production of granular cells was nearly three times greater than the controls, and the increase in cell size was almost twice the control values. Both increased cellular synthetic activity and cell hypertrophy occurred in conjunction with accelerated cell proliferation.

The nuclei of the buccal epithelium of zinc-deficient rats was assessed cytochemically and autoradiographically by Chen.[36] In this study, 28 Simonsen male rats were randomly separated into equal numbers of control and experimental animals. Controls consumed a pelleted diet supplemented with 36 mg zinc per kilogram. Test animals received the same low zinc regimen (1.3 mg zinc per kilogram diet) without added zinc. After 28 study days, the rats were killed. Eight in each group were used for the cytochemical portion; six, for the autoradiographic component. Tissue blocks of buccal mucosa fixed in 10% neutral formalin were washed thoroughly with distilled water, immersed in ammoniacal silver solution, rewashed, dehydrated, and embedded in Araldite. Sections were then prepared for electron microscopy. For the autoradiographic studies, the right buccal mucosa was dissected out from animals given an i.p. injection of either ^{3}H-arginine or ^{3}H-lysine and killed 1 hr later. The mucosa was fixed in formalin and processed for paraffin sections. Three slides from each specimen were used in the preparation of the autoradiographs. The cytochemical analysis showed a decrease in nucleohistones, and the autoradiographs evidenced a decreased incorporation of the labeled amino acids in the zinc-deficient hyperplastic buccal epithelium compared to the controls.

Zinc lack causes parakeratosis, intraepithelial cell death, vesicle formation, and ulceration in the nonkeratinized portion of the oral mucosa of animals rendered deficient by dietary means. To gain further insight into the pathogenesis of these mucosal changes, Chen[37] examined the ultrastructure of the oral mucosa of zinc-deficient rabbits. Five New Zealand white male weanling rabbits were placed on a zinc-poor diet containing 2.5 ppm zinc for 4 weeks. Control animals were fed the identical diet supplemented with 50 ppm zinc. At the end of the study period, the parakeratinized mucosa of the posterior cheek and the ulcerated mucosa of the floor of the mouth were removed from the zinc-deficient animals and processed for electron microscopic examination.

The parakeratinized cells on the epithelial surface had a thickening of the inner leaflet of the plasma membrane, dense packing of tonofilaments, and incomplete disintegration of nucleus and organelles. Intraepithelial necrotizing cells were characterized by disruption and disappearance of plasma membrane, packing of tonofilaments, and degradation of nucleus and organelles. Material covering the ulcerated areas was composed of cell debris, clumps of tonofilaments, neutrophils, microorganisms, and fibrinous exudate. The results indicate that the intraepithelial cell death in zinc deficiency is different from parakeratosis and appears to be a process that precedes vesicle formation and ulceration.

Mesrobian and Shklar[38] looked into the effect of dietary zinc sulfate on the healing of experimental extraction wounds. In this study, 62 Syrian hamsters were subjected to a lower first molar extraction under i.p. anesthesia. Of the group, 31 hamsters received a supplement of zinc sulfate in the drinking water amounting to an average daily intake of 1.3 mg of the salt, whereas 31 were given unsupplemented water. All consumed the same laboratory chow. Sacrifice times ranged from 3 days to 14 weeks after extraction. At necropsy, the jaws were removed, fixed in 10% formalin, and processed for histologic examination of hematoxylin and eosin stained sections.

There was a notable improvement in alveolar bone wound healing in the zinc supplemented hamsters 3 days postoperatively compared to the controls. The inflammatory infiltrate was less dense, and there was more osteoid formation and more rapid epithelialization of the extraction wound sites in the zinc-supplemented animals. After the first postoperative week, both groups healed equally well. The action of the zinc during early wound healing was apparently mediated through metalloenzymes and coenzymes such as alkaline phosphatase, carbonic anhydrase, carboxypeptidase, and lactic acid dehydrogenase, in all of which zinc is an essential component.

Copper Studies

Furstman and Rothman[39] described the effects of copper deficiency on the mandibular joint and alveolar bone of swine. Four weanling pigs were fed a diet deficient in copper to which sodium sulfide was added to decrease the amount of copper absorbed from the gut. Supplemental iron was included to prevent anemia. One control piglet was given the same diet plus copper sulfate. All of the experimental animals remained on the low copper regimen until the serum copper level fell below 10 mg% and the hematocrit below 30. This took approximately 100 days. Following sacrifice the heads were halved midsagittaly, fixed, blocked, dehydrated, and processed for nitrocellulose embedding. Maxillary and mandibular molar regions and the mandibular joint were sectioned in the frontal and sagittal planes and stained with hematoxylin and triosin and with Mallory's connective tissue stain.

The spongiosa of the maxilla and mandible in the copper-deficient pigs was osteoporotic with a marked loss of trabeculae and intact lamina dura. The few remaining trabeculae were surrounded by numerous osteoclasts. Marrow spaces were greatly reduced in size and had lost most of their myeloid cellular elements. In the condyles of the copper-deficient animals, the cartilaginous layer was greatly increased in width and contained several resting lines. Hypertrophied cells were very prominent throughout the lower half of this layer. There was a notable decrease in size and number of trabeculae. They were characterized by a calcified cartilage core covered either by flattened, spindle-shaped cells or by a very thin layer of bone. Condyles of the copper-deficient swine showed a dissociation between chondroblastic and osteoblastic activity since the cartilaginous matrix continued to proliferate and calcify without bone replacement. Condylar changes closely resembled those previously noted in the long bones of copper-deficient pigs.

Iron Studies

Monto et al.[40] studied the gross and microscopic appearance of the tongue, exfoliative cytology, and histology of the lingual mucosa before, during, and after iron therapy in patients with moderately severe classic iron-deficiency anemia. In those with complete atrophy of the lingual mucosa, biopsy specimens disclosed stunted papillae on microscopic examination. Atrophy of the filiform papillae preceded that of the fungiform papillae. After iron therapy, there was a restoration toward a normal state.

Exfoliated squamous epithelial cells of the tongue obtained during severe iron deficiency prepared and stained by the Papanicolaou technic demonstrated a marked reduction in cornified and keratinized population. The cytoplasmic diameter of the exfoliated cells was reduced with paradoxical enlargement of the nucleus. Nuclear pattern was altered, and abnormal cellular maturation was found. This was evidenced by a disturbed nuclear arrangement, increase in nucleoli, presence of double nuclei and karyorrhexis. Histologic sections of the tongue from specimens taken during iron-deficiency glossitis revealed atrophy or absence of papillary formation, lack of keratinization, and thinning of the mucosa. After iron therapy and adequate diet, the cytologic and histologic appearance and the gross structure of the tongue returned to normal.

Iodine Studies

Administration of excessive amounts of iodide to hamsters leads to damage to the epithelial cells lining the distal part of the collecting ducts of the submandibular gland. In contrast, the parotid and sublingual glands are unaffected. To establish whether the variations in response are related to the iodine concentrating capacity of the three major salivary glands, Follis[41] looked into the reparative process of iodine-induced hamster sialadenitis, the effects of certain pharmacologic agents on the gland response, and the reactivity of the salivary glands of other species to excess iodine intake. For the reparative studies, three groups of adult male hamsters were given 20 mg potassium iodide i.p. for 2, 4, or 8 days. On days 3, 5, and 9, the right submandibular gland was excised and examined microscopically. Groups of animals were serially sacrificed 2 to 62 days later, at which time the left gland was removed and processed for microscopic examination. For the pharmacologic studies, animals were first given an injection of either (1) atropine sulfate — 0.5 mg in two doses separated by a 4-hr interval, (2) pilocarpine nitrate — 5 mg on the same schedule as atropine, (3) thiouracil — one 50 mg dose, or (4) potassium thiocyanate — one 50 mg dose. All of the animals received potassium iodide i.p. 10 min after the first drug injection and were sacrificed 24 hr later. For the studies of species differences, rats, mice, rabbits, cats, and dogs were given potassium iodide i.p. at a level of 500 mg/kg body weight.

Administration of 10 to 200 mg potassium iodide to male hamsters weighing approximately 125 g produced a swelling of the interlobular connective tissue in the submandibular glands. Within 2 to 4 hr, the cells lining the distal portion of the ducts became necrotic and the walls infiltrated with inflammatory cells. By 24 hr, the proximal ducts were dilated and the acini atrophic. After 8 days, there was necrosis of the duct epithelium, obstruction of the ducts, and atrophy and inflammation of the acinar elements. The parotid and sublingual glands were spared, indicating that the submandibular glands of the hamsters concentrated the iodide to a greater degree than the other glands. When administration of iodide was stopped, the healing sequence was rapid. By the 10th day, the glands were normal except for the prominence of ducts that continued to be evident even after 2 months. Atropine protected the glands from excess iodide, whereas pilocarpine intensified the damage. Thiouracil and sodium thiocyanate had no apparent effect. A similar iodide-induced sialadenitis of the submandibular gland was found only in mice. The other species tested were not affected.

Fluoride Studies

Weatherall and Weidmann[42] detailed the mandibular changes in chronic fluorosis created experimentally in rabbits and sheep. In this study, 18 adult rabbits were given drinking water containing 500 ppm F *ad libitum* for 3 days to 21 weeks; one sheep was fed a daily capsule of 10 mg F as sodium fluoride until death at 1 month. Numerous exostoses that varied in form, distribution, and extent developed in the mandibles of the test animals. In some, they appeared as isolated deposits; in others, they covered the entire surface.

The rabbits demonstrated an increase in the width of the mandible that resulted principally from exostosis formation in the buccal periosteum. Cellular activity in bone was increased. The periosteum was thickened and rich in proliferating fibroblasts. Resting and active osteoblasts were increased in number. The exostoses consisted largely of coarse bundled woven bone. They usually formed an open trabecular network radiating from the periosteal surface. Radiographic, chemical, and histologic examination showed that the exostotic bone was no more highly calcified than the adjacent cortical bone; often it was less calcified. Bone resorption was a marked but not invariable feature of the fluorotic bone. Wide seams of osteoid in the sheep exostoses suggested a deficiency in calcification.

Schour and Smith[43] analyzed the histologic changes in the incisors of rats given fluoride by injection. In this study, 32 rats, 90 to 270 days of age, received one to eight injections of 0.1 to 0.9 m*l* 2.5% sodium fluoride. In those with multiple injections, the intervening intervals were 24 to 48 hr; in those with single injections, the period between administration and sacrifice was 1 to 48 hr. Incisors of 16 litter mates were used as controls.

Both the enamel and dentin showed a pair of light (disturbed) and dark (recovery) incremental layers after each dose of sodium fluoride. The width of each pair was approximately 32 μm for injections given at 48-hr intervals and approximately 16 μm for those given at 24-hr intervals. The light layers representing the immediate response were imperfect in formation and calcification; the dark layers representing the recovery phase were normal in formation and normal or excessive in calcification. The incremental pattern was a constant finding except in single-injection experiments of 24 hr or less of postinjection life. Disturbances arose when the dosage was increased or when administration was continued for more than five injections at 24-hr intervals. Earliest changes in acute fluorosis were seen in animals given one injection of 0.3 mℓ 2.5% sodium fluoride and allowed to live only 1 hr. They occurred in the ganoblastic layer in the posterior region of the incisors and were denoted by an abnormal character and distribution of globules. The incremental surface of the organic enamel matrix lacked normal arrangement and was covered with hemispherical globules that stained deeply with hematoxylin in animals allowed to live for 12 to 24 hr postinjection. Fluoride apparently exerted a direct local effect on the ameloblasts in these experiments.

There are growing indications that fluoride may be of value in the prevention and treatment of some of the demineralizing bone diseases. The prevalence of osteoporosis has been shown to be inversely related to the fluoride content of the water supply. Treatment of osteoporotic patients with sodium fluoride has resulted in skeletal recalcification, conversion from a negative to positive calcium balance, and reduction in renal calcium excretion.

To test the demineralizing-resistant properties of fluoride, Levy et al.[44] investigated the effect of parathyroid hormone on the alveolar bone of marmosets pretreated with fluoridated and nonfluoridated drinking water. In this study, 44 healthy, acclimated young cotton ear marmosets (*Callithrix jacchus*) were fed a diet containing all of the known nutritional essentials for this primate. Of the group, 20 animals (Group I) were restricted to drinking water containing 50 ppm F and 24 (Group II) to nonfluoridated drinking water available *ad libitum*. After 5 months on this regimen, 12 animals in Group I and 10 in Group II were given parathyroid hormone in amounts ranging from 300 to 800 USP units. The hormone preparation was administered in divided doses over a 3-day period. Each dose contained 100 USP units parathyroid suspended in 0.5 mℓ 16% gelatin. Marmosets not given the parathyroid served as group controls. All were killed with ether and necropsied the day after the last parathyroid injection. The skulls were fixed in neutral formalin and split midsagittally. They were cut into tooth-containing blocks, decalcified in formic acid, processed in paraffin, sectioned, and stained with hematoxylin and eosin, Masson's trichrome, PAS, and Weigert's iron hematoxylin.

Administration of parathyroid hormone to animals not pretreated with fluoride resulted in rapid and severe bone loss that affected all parts of the alveolar process. The severity of the response was, in general, dose related. In all instances, there was a decided upsurge in osteoclastic activity denoted by the appearance of numerous multinucleated giant cell osteoclasts, Howship's lacunae, and resorption of the trabeculae and of the cortical and cribriform plates. Resorption of the plates followed a centripetal course with the osteoclasts almost invariably located on the marrow side of the involved bone. The intense osteoclastic activity was paralleled by a proliferation of young fibroblasts in the marrow spaces and culminated in a thinning and disappearance of the trabeculae and in diminished thickness and breaks in the continuity of the plates of compact bone (Figures 1A and 1B).

Alveolar bone patterns of marmosets given parathyroid hormone after pretreatment with fluoride showed substantially less hormone-induced bone resorption than in animals not given fluoridated water and injected with identical amounts of hormone. At levels up to 600 USP units of parathyroid, bone loss in the fluoride-supplemented group was minimal and confined largely to a few isolated areas of osteoclasis. At 800 USP units, osteoclastic activity was more widespread, but still considerably less than in the animals on the nonfluoridated water. Fluoride-precipitated changes in bone structure and bone metabolism may thus have a sparing effect against the lytic action of parathyroid hormone excess (Figures 2A and 2B).

Both deficiencies and excesses of various dietary minerals have been found to markedly affect the oral structures of laboratory animals. The ease with which such states can be created in animals whose life cycles are so compressed in time compared to

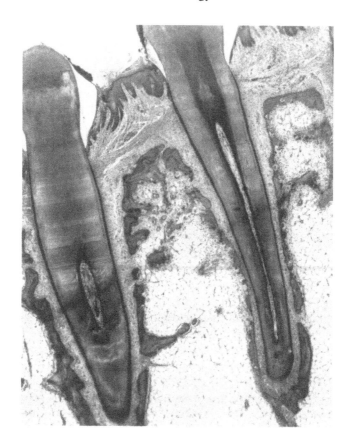

FIGURE 1A. Maxillary premolar area from nonfluoride treated
marmoset given 600 USP units parathyroid hormone showing extensive osteoclastic resorption and bone marrow fibrosis. (PAS and hematoxylin stain; magnification × 75.) (From Levy, B. M., Dreizen,
S., Bernick, S., and Hampton, J. K., Jr., *J. Dent. Res.,* 49, 816, 1970.
With permission.)

man has provided an abundance of supportive data. Calcium, phosphorus, magnesium, and copper have been found to be required for normal tooth and/or bone formation, fluoride for bone and tooth maintenance, and zinc and iron for maturation
of the epithelial lining of the oral mucosa. Further progress in this area awaits the
refinement of technics that will permit complete elimination of even the most infinites-
imal traces of the trace minerals.

VITAMIN DEFICIENCY

Many of the vitamin deficiencies produce tell-tale signs and symptoms in the mouth.
These are represented by changes in sensitivity, salivation, and taste acuity and by
alterations in the color, continuity, and topography of the mucosa. Most of the clinical
manifestations are subject centered. Patients followed through many recurrences of
the same vitamin-deficiency state develop oral lesions in the same regular order and in
the same definitive site in a highly predictable and programmed manner. As demonstrated below, experimental stomatology has made major contributions to the unveil-
ing of the oral reflections of avitaminoses.

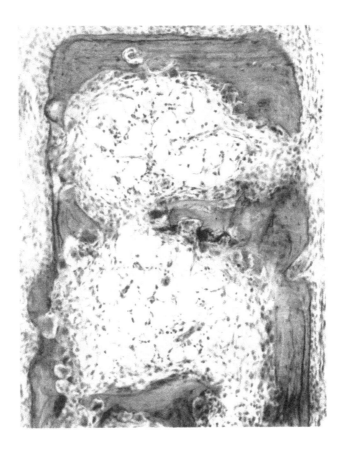

FIGURE 1B. Higher magnification of alveolar crest area shown in (A). Note numerous multinucleated osteoclasts and young fibroblasts. (PAS and hematoxylin stain; magnification × 250.) (From Levy, B. M., Dreizen, S., Bernick, S., and Hampton, J. K., Jr., *J. Dent. Res.*, 49, 816, 1970. With permission.)

Vitamin A Studies in Animals

Both a deficiency and an excess of vitamin A will produce marked effects on the mouth structures of experimental animals. Burn et al.[45] described the sequence of gross and microscopic changes in the incisor and molar teeth of rats maintained for up to 1 year in a state of chronic vitamin A deficiency by feeding minimal quantities of the vitamin. Young rats from litters of six to eight were restricted to a diet of dextrin, vitamin A-free casein, salts, and supplements of brewer's yeast and vitamin D at weaning. To hasten the depletion of vitamin A, they were removed from the mothers during the nursing period. At the appearance of the first definite signs of vitamin A deficiency denoted by growth cessation, weight decline, photophobia, or predominance of cornified cells in vaginal smears, they were given small quantities of vitamin A by mouth, the amount determined largely by clinical appearance. Control rats were given the identical diet plus a daily dose of 300 IU vitamin A in the form of cod liver oil. Groups of five test and control animals were killed at intervals ranging from 15 to 365 days. At necropsy, the jaws were excised, cut into blocks and prepared for histologic examination. Sections were stained with hematoxylin and eosin and Masson's.

Striking gross and microscopic derangements were found in the incisor teeth of the rats with chronic vitamin A deficiency. There was a progressive loss of pigmentation and translucency, twisting of the incisors, transverse and longitudinal ridging, pulp

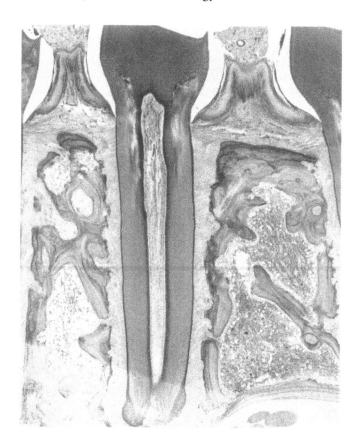

FIGURE 2A.   Maxillary premolar area from fluoride-treated marmoset given 600 USP units of parathyroid hormone. (PAS and hematoxylin stain; magnification × 75.) (From Levy, B. M., Dreizen, S., Bernick, S., and Hampton, J. K., Jr., *J. Dent. Res.,* 49, 816, 1970. With permission.)

exposure, and eventual exfoliation of the erupted portion. Older animals developed large odontomas.

Microscopic changes were present in the ameloblasts, odontoblasts, and enamel organ. There were focal areas of degeneration in the ameloblastic layer, atrophy and complete disappearance of the lingual odontoblasts, and enlargement and hyperplasia of the labial odontoblasts. The lingual dentin was resorbed and eventually lost, permitting embryonic pulp cells to grow unrestrained into the surrounding soft tissue and form tumor masses.

Odontomas were found in 62% of the animals of advanced age. The odontomas were characterized by an embryonic type of connective tissue with inclusions of odontoblasts, epithelial cells, keratin, dentin, and newly formed germinal centers. The molar teeth showed no gross changes. Microscopically, they contained intensely stained dentin and enamel remains and showed focal resorption of cementum and roots.

Schour et al.[46] investigated the influence of both vitamin A deficiency and repletion on the incisor teeth of the albino rat. In this study, 199 white rats were placed on a vitamin A-deficient diet beginning at weaning and extending for 9 to 81 days. Of the group, 84 were given various types of vitamin A replacement therapy, and 95 were subjected to vital staining with alizarin red S to study rates of dentin apposition. After death, the heads were decapitated, fixed in formalin, split midsagittally and radi-

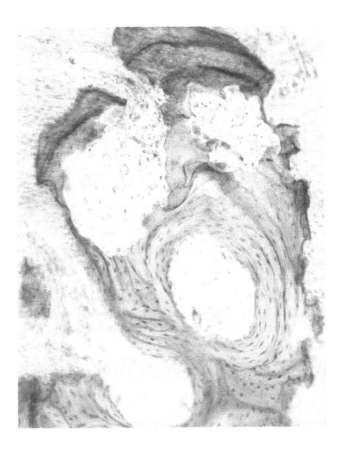

FIGURE 2B. Higher magnification of alveolar crest area shown in (A). Note isolated focus of osteoclastic activity. (PAS and hematoxylin stain; magnification × 250.) (From Levy, B. M., Dreizen, S., Bernick, S., and Hampton, J. K., Jr., *J. Dent. Res.*, 49, 816, 1970. With permission.)

ographed. The jaws were decalcified, dehydrated, celloidin-embedded, sectioned, and stained with hematoxylin and eosin.

Teeth of animals given injections of the dye were examined in ground and in decalcified sections. Measurements were made with a micrometer eyepiece. Radiographically, the incisors showed striking distortions. The convex border of enamel-covered dentin was widened, whereas the concave border representing lingual dentin was absent in the proximal third and thinned in the middle third.

Histologically, the primary effect of the vitamin A deficiency was on histodifferentiation of the odontogenic epithelium. The lingual odontogenic epithelium was disturbed and incomplete. There were morphologic and functional alterations in the lingual odontoblasts, and the lingual dentin was abnormally thin. Along with the lack of histodifferentiation, there was a continuation of the proliferative activity of the odontogenic epithelium leading to pulp invasion by epithelial cords. The rate of dentin apposition was selectively altered, with the enamel-covered dentin undergoing acceleration and the cementum-covered dentin deceleration. Replacement therapy resulted in resumption of the normal rate of dentin apposition and prompt differentiation of peripheral pulp cells into odontoblasts.

That vitamin A deficiency is the most important vitamin deficiency affecting tooth formation was clearly demonstrated by Wolbach and Howe,[47] who found that the initial effect of vitamin A lack in both rats and guinea pigs was on the enamel organ of the incisor teeth. In each rodent, the earliest response was atrophy of the ameloblasts followed by atrophy of the rest of the enamel organ and of the odontoblasts. The deficiency also produced calcification of the enamel organ in the rat and ossification of the enamel organ in the guinea pig.

Moore and Mitchell[48] found that the incisor teeth of albino and piebald rats kept on a diet deficient in vitamin A underwent depigmentation and a decline in iron content. The teeth were examined by observing their color and by analyzing the enamel for iron by flame spectrophotometry. The fall in the iron level of the enamel and the depigmentation occurred in both strains of rats. On realimentation with vitamin A, increases in the pigmentation of the bleached enamel did not occur for several weeks.

The histopathology of the major and minor salivary glands in rats maintained on a diet deficient in vitamin A for approximately 50 days has been detailed by Trowbridge.[49] In this study, 32 Long-Evans rats, 19 days of age, were divided into experimental and control groups and pair-fed a purified vitamin A-free diet. The regimen was supplemented with 22,000 IU vitamin A per kilogram for the control group. Test animals were sacrificed along with their paired controls at the earliest display of acute signs of vitamin A deficiency. They were considered acutely deficient when there was blanching of the incisors, xerophthalmia, and arrested or markedly depressed growth.

The principal effects of the deficiency were acinar atrophy, proliferation and metaplasia of the ductal epithelium, and distention of the ducts. Obstruction of the excretory ducts was a frequent occurrence due to the formation of keratinizing epithelium. Inflammatory changes were related either to degenerative alterations in the ducts or to infection of the glands. Marked reduction in activity of the oxidative enzymes, lactic and succinic acid dehydrogenase were detected during transformation of the duct cells to keratinizing epithelium. Metaplasia was accompanied by pronounced mitotic activity and proliferation of squamous cells in the ductal lamina. Submaxillary glands were almost completely lacking in convoluted granular tubules. The sublingual glands and lingual mucous glands were not affected by the deficiency to the same degree as the submaxillary and parotid salivary glands and the serous glands of the tongue. All experimental animals developed abscesses in the sublingual salivary glands.

The effects of vitamin A deficiency on the oral structures of the hamster were elicited by Salley and Bryson.[50] In that study, 50 weanling albino hamsters, 20 days of age, from five separate litters, were distributed into three groups. Group I (22 animals) was fed a diet of known composition deficient in vitamin A; Group II (12 animals) was given the same diet supplemented with 250 IU vitamin A in cottonseed oil administered by mouth; Group III (16 animals) was provided with Purina® laboratory chow and served as controls. At the end of 15 study weeks, all survivors in Group I and all animals in Group II and III were sacrificed by chloroform inhalation. Livers were removed *in toto* and analyzed for vitamin A content. The right mandible, tongue, major salivary glands, trachea, lung, spleen, pancreas, and ovaries or testes were formalin-fixed and prepared for histologic examination.

All groups showed a similar weight gain until the 8th week when the vitamin A-deficient animals experienced a decided weight loss and thereafter followed a continuous downhill course. Microscopically, the most striking changes were exhibited by the major and minor salivary glands. Serous acini showed the earliest and most severe alterations. Mucous acini were involved to a lesser degree. The principal effect was squamous metaplasia of the epithelium. The glands were also markedly infiltrated by inflammatory cells. Duct obstruction and cyst formation were present in areas of squamous metaplasia with keratinization. Odontogenic tissues underwent squamous metaplasia of the ameloblastic layer at the posterior end of the incisor teeth and irregular dentin formation.

There is a general agreement that vitamin A is essential for maintaining the integrity of the lining epithelium in many parts of the body although the precise mechanism is obscure. Oral epithelial responses to a lack of vitamin A have been studied in rats, guinea pigs, hamsters, dogs, cattle, and rhesus monkeys. In each of these species, vitamin A deficiency produces keratinizing metaplasia of the normally columnar ductal epithelium in the major and minor salivary glands.

Since the nutritional adequacy of the experimental diets used in some of the earlier studies is suspect, the oral histopathologic reactions to avitaminosis A in nonhuman primates was reexamined by Dreizen et al.[51] Four adult, healthy, colony-conditioned cotton top marmosets (2 male, 2 female) (*Saguinus oedipus*) were confined to a chemically identified diet devoid of vitamin A and carotene. Four other animals (2 male, 2 female) given the identical diet plus daily supplements of 360 IU vitamin A acetate acted as controls. The monkeys were kept on the vitamin A-free diet until they became moribund and were killed after 67, 106, 127, and 132 study weeks by cardiac puncture exsanguination. Assigned controls were sacrificed at the same time. All structures removed at necropsy were fixed in 10% buffered formalin. The heads were sawed into blocks, demineralized, and processed for histologic examination. Sections were stained with hematoxylin and erythrosin B, Masson's trichrome, and Weigert's iron hematoxylin.

Each of the vitamin A-deprived marmosets had pathognomonic microscopic markers of avitaminosis A in the salivary glands, lingual glands, and genitourinary and respiratory tracts. Histologically, these included substitution of the specialized lining epithelium in parts of the genitourinary and respiratory systems by keratinized stratified squamous epithelium and keratinizing metaplasia of the ductal epithelium of the parotid, submaxillary, sublingual salivary glands and lingual glands. There was also aberrant stratified squamous epithelialization of the lamina propria in the gingiva, demonstrating that both the specialized and nonspecialized oral epithelium are sensitive to vitamin A deficiency and differ mainly in the expression of the deficiency response (Figures 3A and 3B).

Hayes et al.[52] noted the morphologic alterations in the fine structure of the parotid ducts of vitamin A-deficient calves. Six male Holstein calves were restricted to a vitamin A-deficient diet starting at 35 days of age. They were later assigned to either a deficient regimen (8 μg vitamin A per kilogram diet per day) or to an adequate daily supply of vitamin A acetate (108 μg/kg diet per day). After a 25-week comparison period, the animals were anesthetized with sodium pentothal, the heads perfused with glutaraldehyde and processed for electron microscope examination.

Control animals appeared to have two morphologic cell types in the parotid duct epithelium, mucoid and fibrous. The mucoid cells were characterized by cisternal profiles of granular reticulum, extensive Golgi membranes, and dense aggregates of intracytoplasmic mucin. Fibrous cells did not actively synthesize mucus, but were typified by cytoplasmic tonofibrils and a scarcity of granular endoplasmic reticulum and Golgi membranes. In the deficient epithelium, mucoid cells seldom developed; fibrous cells predominated without evidence of keratinization. These findings suggest that vitamin A deficiency prevents activation or differentiation of mucus-producing cells. Cell populations most affected by the deficiency are those subject to rapid turnover that can differentiate in more than one direction.

Postextraction wound healing in vitamin A-deficient hamsters was investigated by Hirschi.[53] In this study, 40 weanling hamsters 20 days of age were divided into (1) a control group of 8 fed a standard dog biscuit diet and (2) an experimental group of 32 given vitamin A test diet USP Number 12. On the 40th experimental day, the lower left first molars were extracted under i.p. barbiturate anesthesia. The hamsters were subsequently sacrificed at intervals up to 20 days, necropsied, and the wound sites prepared for histologic survey.

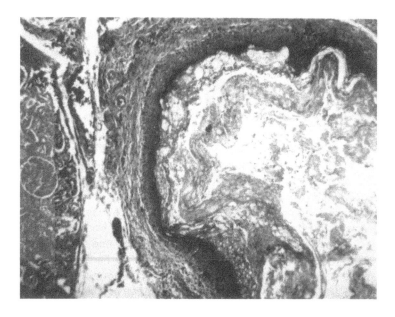

FIGURE 3A.   Ureter of vitamin A-deficient marmoset demonstrating replacement of transitional epithelium by keratinizing stratified squamous epithelium. (Hematoxylin and erythrosin B stain; magnification × 20.) (From Dreizen, S., Levy, B. M., and Bernick, S., *J. Dent. Res.*, 52, 803, 1973. With permission.)

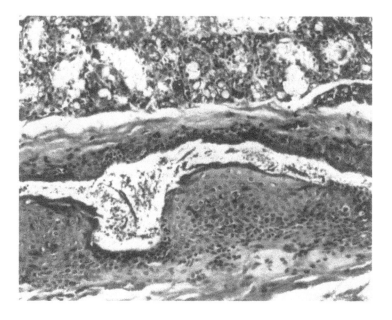

FIGURE 3B.   Keratinizing metaplasia of epithelial cell lining of submaxillary gland duct in vitamin A-deficient marmoset. (Hematoxylin and erythrosin B stain; magnification × 50.) (From Dreizen, S., Levy, B. M., and Bernick, S., *J. Dent. Res.*, 52, 803, 1973. With Permission.)

The vitamin A-deficient animals showed a loss of their golden color by day 14 and xerophthalmia by day 28. Hemorrhage from the genitalia and bowel developed during the later stages of the deficiency. Although new bone formation in the healing sockets was initiated earlier in the deficient animals, the quantity formed was decidedly lower than in the controls. Animals with avitaminosis A also showed an extensive proliferation of bone marrow cells, infiltration of acute and chronic inflammatory cells in the alveolar bone, and a decrease in quantity of old and new bone.

To clarify the role of osteoblasts and osteoclasts in vitamin A deficiency, Frandsen and Becks[54] studied the effects of hypovitaminosis A on bone healing and endochondral ossification in rats. In this study, 45 male Long-Evans rats were assigned to a group of 18 started on a vitamin A-free diet at 14 days of age, a group of 18 pair-fed a control diet, and a group of 9 given the control diet *ad libitum.* The left lower first molar was extracted from each animal at 58 to 62 days old. Following extraction, the rats were killed at intervals ranging from 7 hr to 46 days. Heads and tibias were removed and fixed in 10% formalin, radiographed, and prepared for histologic examination.

After 25 to 30 days on the deficient diet the rats exhibited evidence of peripheral nerve degeneration and xerophthalmia. By 40 days, they were getting moribund and had to be fed suboptimal maintenance doses of vitamin A (3 μg vitamin A alcohol in oil daily) at intervals to assure survival until the end of the experiment. The main histologic changes associated with the avitaminotic state were (1) decreased osteoblastic activity evidenced by retardation in formation of immature bone in the healing tooth socket, (2) suppression or absence of osteoclastic activity manifested by failure of remodeling resorption of immature bone in the tooth socket and tibial shaft, and (3) the immature bone in the vitamin A-deficient rats contained more and larger osteocytes than in the controls with distinctly eosinophilic cytoplasm.

The first experimental evidence that a deficiency of a single vitamin in the diet of a pregnant animal could result in deformities in the offspring was demonstrated by Hale.[55] Feeding pregnant sows a regimen lacking in vitamin A produced progeny with numerous malformations. Included were cleft lip, cleft palate, anophthalmia, microphthalmia, accessory ears, and malformed hind legs. The teratologic effect of a maternal deficiency of vitamin A stems from the need of the developing embryo for the vitamin at critical developmental stages in the orderly process of normal differentiation.

That an excess of vitamin A in the diet of pregnant animals can also induce congenital anomalies was demonstrated conclusively by Cohlan.[56] Female Wistar rats weighing 150 to 200 g were mated with males by exposure in the preestrous state overnight. The morning of exposure was considered the first day of gestation. Experimental groups were fed an aqueous preparation of vitamin A in doses from 15,000 to 75,000 IU for varying periods during gestation. These ranged from a single dose on 1 day to daily doses from the 2nd to 16th day, inclusive. Controls were fed an equal volume of diluent. All feedings were by intubation. Pregnancy was allowed to proceed, and in most instances the animals were killed on the 20th to 21st day at or near term.

Only 25 of the 210 females fed 35,000 IU daily from the 2nd, 3rd, or 4th to the 16th day of gestation produced litters of 148 newborn. The successful pregnancy rate was 12% compared to 84% for the controls. This marked litter failure in the vitamin A-fed group was due to fetal death and resorption. Of the 148 offspring, 77 (52%) exhibited gross anomalies of the skull and brain. Anomalies such as cleft palate, shortening of the mandible and maxilla, gross eye defects, and hydrocephalus occurred either alone or in association with the major cranial deformity. No gross congenital malformations were found in the control group.

The palates of the offspring from the vitamin A-fed mothers had triangular shaped defects with the apex extending anteriorly to the incisive papilla but not involving the alveolar process or lips. This anomaly appeared to be identical to that described in vitamin A deficiency and was believed to be due to a failure of fusion of the palatine process of the maxilla and the horizontal process of the palatine bone. Among the skeletal defects was a foreshortening of the mandible and maxilla that usually involved the corpus of the mandible, but rarely the ramus. A dosage between 15,000 and 25,000 IU seemed to be the minimal daily dose of vitamin A needed to produce the gross malformations. These findings suggest that the excess vitamin A seemingly exerts a teratogenic action directly on the developing embryo rather than indirectly through maternal stress.

Kalter[57] used the method of maternal overdosage with vitamin A to produce orofacial anomalies in inbred mice. Young adult A/Jax, DBA/1 Jax, and C3H/Jax inbred females were placed with males of their own strain, one pair to a cage, and observed each morning for a copulation plug. The presence of a plug indicated ⅓ day after copulation. Impregnated females were placed in individual cages and given one oral dose of 10,000 IU vitamin A 8 ⅓ and 9 ⅓ days after conception. All were fed Purina® lab chow and fresh tap water *ad libitum* throughout pregnancy. They were sacrificed 17 ⅓ days after conception or approximately 2 days before term. Resorption sites were counted, and fetuses were removed and examined microscopically before fixation in Bouin's solution for serial sectioning or in 95% alcohol in preparation for skeletal staining with alizarin S.

There were no strain differences in the type or frequency of experimentally induced congenital anomalies. Of the 152 fetuses that survived, 62.5% had dermal tabs and hillocks of varying length and size at the corners of the mouth, 33.6% had microstomia, 17.8% micrognathia, 11.2% shortening of the upper jaw, and 2.0% median cleft of the mandible. Of the 112 fetuses that could be examined in various ways, 70 (62.5%) had cleft palates.

Deuschle et al.[58] confirmed the effectiveness of maternal hypervitaminosis A as a teratogenic method by administering single doses of 75,000 and 150,000 IU of vitamin A orally to pregnant rats on days 9, 10, and 11 of gestation. Eight fetuses with large protruding eyes were serially sectioned. The outstanding oculodentofacial abnormalities in these specimens were exophthalmus, maxillomandibular ankylosis, presence of heteroptic cartilage in the maxilla, and absence of some molar teeth. The total syndrome was manifested chiefly in derivatives of the mandibular arch, with the eye apparently secondarily involved because of the regional relationship. Both mesodermal and ectodermal structures were affected.

Abramovich[59] compared the cleft palate inducing capacity of hypervitaminosis A with *Lathyrus odoratus* in the rat. In this study, 32 female white rats were mated and the date of conception established. They were then separated into four groups of eight animals each. Group I received a standard diet and was used as a control; Group II was fed a diet containing 50% *L. odoratus* ; Group III was injected with 100,000 IU vitamin A on the 10th and 11th day of pregnancy. All fetuses were removed by laparotomy on the 18th day of gestation. They were fixed in formalin, embedded in paraffin, and sectioned frontally. Micrometric measurements were made in the anterior region in a plane that passed through Jacobson's organ; in the posterior region, in a plane that passed through the first molar tooth germs. Measurements were made of the height of the head, height of the nasal septum, and height of the cartilage in the nasal septum in the anterior region, and of the width of the tongue at the lingual groove and of the height of the tongue from the dorsum to the geniohyoid muscle in the posterior region.

Both agents produced cleft palates in almost all of the fetuses. *L. odoratus* inhibited development of the palatal shelves and caused the appearance of cartilaginous masses at the level of the first mandibular molar; hypervitaminosis A induced agenesis of the tooth germs, inhibited formation of the palatal shelves, and caused cartilage to appear in the region of the maxilla. The nasal septum cartilage was significantly reduced in height, and the mandibles were significantly thicker in the experimental groups than in the controls, although the heights of the face and of the total nasal septum were similar. The results suggest that cleft palate formation is related more to lack of development of the inferior portion of the chondrocranium than to increase in mandibular thickness.

Cavalaris and Krikos[60] produced mucous metaplasia of the normal keratinizing epithelium of the hamster cheek pouch by continuous topical application of high levels of vitamin A. In this study, 20 male golden Syrian hamsters 3 to 4 months of age weighing 100 to 125 g, were anesthetized with i.p. pentobarbital. Cheek pouch areas from the commissures of the lips to the shoulder girdle were shaved bilaterally and the pouches everted and cleaned with cotton swabs and distilled water. Each pouch was drawn through an undermining incision made posterior to the commissure. A paraffin rod containing 50,000 IU vitamin A was placed in the right pouch through the oral cavity and the pouch looped with a silk suture to entrap the rod in position. The right pouch served as the test site; the left, as a control for the possibility of systemic effects.

Topical application of vitamin A to the hamster cheek pouch caused a spectrum of epithelial changes ranging from hyperplasia to focal areas of mucous metaplasia. The metaplastic changes were related to the duration of treatment. Pouches treated for 5 days underwent hyperplasia, parakeratosis, and necrosis. Mucous metaplasia appeared after the 10th day of treatment and became progressively more pronounced with time. There were no vitamin A-related changes in the control pouches, indicating that the action of vitamin A activity was local rather than systemic.

Hypervitaminosis A has also been found to cause hypertrophy of the tongue muscles in the rat. Lopes et al.[61] divided 20 male Wistar rats weighing approximately 85 g into two groups of 10 each. Group I was injected i.p. with 200 IU vitamin A per gram body weight daily for 10 days. Group II was similarly injected with saline. Both groups were fed a balanced diet and drinking water *ad libitum.* The rats were sacrificed by ether inhalation and the tongues dissected out and fixed in Bouin's solution. After fixation, they were processed for histologic examination of hematoxylin and eosin and of Masson's trichrome stained sections.

Generalized hypertrophy of the tongue muscles was found in all of the test animals. These observations were confirmed by morphometric technics. Increased cytoplasmic and decreased nuclear volumes were demonstrated by the Chalkley method, karyometry, and microscopy. The changes were attributed to the direct action of vitamin A on the organelles of the muscle fibers.

Lopes et al.[62] investigated the effects of hypervitaminosis A on the morphology of the rat sublingual salivary glands. In this study, 20 male rats weighing about 100 g were separated into two groups of equal size. Group I was given i.p. injections of 200 IU vitamin A per gram body weight daily for 12 days; Group II received comparable injections of saline. Upon development of hypervitaminosis A, the test rats were anesthetized and the sublingual glands surgically excised and fixed immediately in alcohol-formalin-acetic acid solution. Glands from the control animals were treated in identical fashion. After histologic processing, sections were stained with PAS for neutral mucopolysaccharides and with Alcian blue for carboxylated and sulfated acid mucopolysaccharides. Hypervitaminosis A produced a diminution in the size of the acinar and striated granular duct cells, a decrease in cytoplasmic basophilia, and a decrease in the content of neutral and acid mucopolysaccharides in the sublingual salivary gland. These findings were similar to those previously noted in the rat submandibular gland.

Gorlin and Chaudhry[63] reported on the effects of hypervitaminosis A on the teeth, periodontium, and temperomandibular joint of the rat. In this study, 60 Sprague-Dawley weanling male rats were assigned to experimental (40 animals) and control (20 animals) groups. Vitamin A concentrate in corn oil was administered orally to the experimental group each day at a level of 600 IU/g of body weight. A corresponding amount of corn oil was given to the controls. Both groups were fed Purina® laboratory chow *ad libitum*. Animals appearing terminally ill were killed by chloroform anesthesia and necropsied. The head, long bones, and costochondral junctions were fixed in 10% formalin. Bones and teeth were decalcified in formic acid-sodium citrate and the skull split longitudinally. The tissues were embedded in paraffin, sectioned, and stained with hematoxylin and eosin.

By day 9 the experimental animals appeared weak and scruffy. Perioral alopecia and periorbital edema were common. Some had frank exophthalmos. Limping was observed by the 9th to the 14th day, and fractures of the hind legs productive of paralysis by days 16 to 19. Hypervitaminosis A had far less effect on the teeth than on the bones where the most marked changes were in the temperomandibular joint. Condylar head size was greatly reduced and architecture altered. Subperiosteal hemorrhage was present in 78% of the experimental group. The cartilage cap was absent, and there were numerous fractures of the bony trabeculae. One animal had a complete pathologic fracture of the ramus. Marked vascularity of the pulp and periodontal ligament appeared to be the only specific dental changes attributable to the hypervitaminotic state.

Vitamin A Studies in Man

Dinnerman[64] examined the effects of vitamin A deficiency in the unerupted teeth of infants. The study material was comprised of mandibles obtained at autopsy from five infants aged 3 to 7 months with naturally occurring avitaminosis A. The mandibles were fixed in 10% formalin, decalcified in nitric acid, and blocks containing unerupted teeth were embedded in celloidin. Sections were stained with hematoxylin and eosin.

The specialized enamel-forming cells were replaced by a nonspecialized stratified squamous epithelium, and there was atrophy of the enamel organs. Hyperplasia of the enamel, manifested as tears, canal-lined defects, or deeply stained interprismatic substance was found in almost every case. Deposits of atypical enamel on the tooth surface were present in some. In several instances, Hertwig's sheath was completely atrophied. Vitamin A deficiency during the period of active development and calcification of human teeth produced degenerative changes in the ameloblasts that resulted in enamel hypoplasia.

The action of high doses of vitamin A on oral leukoplakia was assessed by Silverman et al.[65] Five adults with oral leukoplakia were given 600,000 IU vitamin A in the form of eight 75,000 IU dissolvable oral troches daily for 4 weeks. Histologic examination of biopsy specimens disclosed a decrease in the thickness of the stratum corneum and reduced degree of cornification. Exfoliative cytology analyses revealed diminished cytoplasmic acidophilia and denucleation. The microscopic changes were accompanied by clinical remissions. Toxic complications consisting mainly of dryness, itching, and scaling of the skin were evidenced by all patients. These disappeared after vitamin A administration was discontinued.

Vitamin B Complex Studies
Thiamine Studies

Abrams[66] elicited the thiamine deficiency manifestations in the neuromuscular components of rat tongue. In this study, 24 male Wistar rats, 90 days of age, were allocated to four groups of 6. Group I was placed on a diet of Purina® rat chow; Group II, a commercial thiamine deficient diet; Group III, the thiamine deficient diet plus salivary gland duct ligation; Group IV, the control diet and duct ligation. Food and water were provided *ad libitum.* One rat from each group was sacrificed weekly. The tongues were removed and fixed in 10% formalin, processed, and sections stained with hematoxylin and eosin, Weil-Weigert's stain for myelin, and Bodian's stain for nerve fibers.

There were no gross differences in the appearance of the tongue between the test and control animals. Histologically, thiamine deficiency with or without salivary gland duct ligation, precipitated changes in both the nerves and muscles of the tongue. The perineurium was disrupted, many neurilemma tubes were empty of myelin, axis cylinders were fragmented, and Schwann cells manifested dark staining of the nuclei. Striated muscles became atrophied and lost their banding. The sarcolemma was disorganized, and muscle fiber nuclei became pyknotic. In addition to the neuromuscular damage in the tongue, the deficient animals displayed marked growth retardation and a change in fur color.

Perri et al.[67] tested the action of some thiamine antagonists on the taste receptors of the frog. The tongue, jowl, suprahyoid region, and large vessels were carefully dissected from pithed frogs (*Rana esculenta*) weighing 20 to 30 g. The preparation was placed in a perspex chamber where the tongue and other parts were arranged in separate cells filled with Ringer's solution. The glossopharyngeal nerve was carefully isolated on the same side and gently sucked up into a glass pipette provided with two platinum leading-off electrodes.

All tongue preparations were initially perfused with Tyrode solution until a steady response was obtained to a test stimulus. Then perfusion with Tyrode solution containing the proper amounts of thiamine, pyrithiamine, and oxythiamine was started. Every 15 min, the tongue was stimulated by applying a calcium chloride solution to the surface for 30 sec. This period was short enough to avoid the onset of significant adaptation by the receptors. Between successive stimulations, the tongue was immersed in Ringer's solution.

Receptor response or the number of action potentials that led off from the glossopharyngeal nerve of the perfused side of the tongue over a 30-sec period of chemical stimulation was recorded on an electronic counter after conventional amplification. Pyrithiamine strongly depressed tongue receptor response to chemical stimulation. The action was almost completely preventable by equimolar amounts of thiamine. Oxythiamine did not affect taste receptor function even when perfused in high concentrations. The results were interpreted as proof that thiamine was essential in frog taste receptor activation. Pyrithiamine interfered with the phosphorylation of thiamine blocking taste receptor function. Lack of action of oxythiamine in this system stemmed from inability to be phosphorylated in vivo and act as a cocarboxylase antagonist.

Riboflavin Studies in Animals

Afonsky[68] described the oral manifestations of riboflavin deficiency in dogs fed a synthetic diet devoid of the vitamin. The regimen was composed of vitamin free-casein, sucrose, hydrogenated vegetable oil, salt mixture, bone ash, and vitamin supplements from which riboflavin was withheld. Three dogs between 1 and 2 years of age were used for this study. The experimental period lasted from 167 to 500 days, during which the animals experienced several attacks of acute deficiency. During each bout, riboflavin was administered and the response observed. Biopsy specimens of the tongue were taken during the depletion and repletion phases. Serial sections were prepared and stained with hematoxylin and eosin, PAS, and Mallory's trichrome.

Lingual lesions developed in all dogs with riboflavin deficiency. Degree of involvement paralleled the severity of the deficiency and the number of acute attacks experienced. Grossly, the glossitis was represented by small, irregularly shaped areas lacking papillae scattered over the anterior two thirds of the dorsum. The denuded patches were most numerous in the middle third. Smaller lesions were detectable only with a magnification lens (magnification × 5). Histologically, there was an atrophic degeneration of both the filiform and fungiform papillae accompanied by a gradual diminution in size. With shrinkage of the connective tissue cores, the papillae lost their shape and formed elevations on the dorsum. These were subsequently extinguished, leaving a smooth surface. Epithelial degeneration started in the basal cell layer and developed into rarefaction of the epithelium and adjoining connective tissue, followed by necrosis and removal of the epithelial debris by phagocytic cells. The degenerative process resulted in a decrease in keratinization. In the final stages, the patches of papillary atrophy were surfaced by a layer of parakeratotic cells.

Levy[69] determined the effect of a diet deficient in riboflavin on the growth and development of the mandibular condyle in mice. In this study, 48 weanling male and female C57 black mice were fed a synthetic diet lacking riboflavin for 1 to 10 weeks; 8 of the mice were refed a stock diet after 4 weeks on the riboflavin-free ration. At the end of each experimental period, the animals were sacrificed and the heads prepared for histologic survey.

Long-term consumption of the riboflavin-deficient diet caused a notable retardation in the growth of the mandible. The condylar growth cartilage became markedly atrophic and narrow. The cartilage cells were much smaller than usual, and osteoblasts were sparse. All changes were reversed and cartilage width was restored upon refeeding with an adequate diet.

Waisman[70] produced a riboflavin-deficiency state in rhesus monkeys *(Macaca mulatta)* by restricting newly obtained animals to a highly purified diet of sucrose, vitamin free-casein, salts, corn oil, cod liver oil, and adequate quantities of pure vitamins except riboflavin. These animals evidenced growth failure after 6 to 8 weeks on the diet. The more significant signs of riboflavin deficiency developed gradually. These included a freckled type of dermatitis that emerged first on the face and groin. Small red dry spots became increasingly prominent, large, and scabby. Administration of 50 mg riboflavin per day brought about a remission of the facial dermatitis within 1 to 3 weeks.

As with vitamin A, maternal deficiencies of riboflavin during pregnancy have been found to cause congenital malformations in the rat. Warkany and Schraffenberger[71] placed 21 female Sprague-Dawley 3-month-old rats that had been reared on an adequate diet on a purified regimen of sucrose, vitamin test casein, vegetable oil, and salt mixture supplemented with thiamine, pyridoxine, pantothenate, nicotinamide, choline chloride, and vitamin A. During the first 2 weeks, the rats maintained weight or had slight gains. Estrous cycles were regular. They were then mated to normal males of the same strain. All but two of the females became pregnant within 3 weeks.

Of the 19 pregnancies, 11 ended in resorption and 8 in delivery of litters at term. A total of 74 young were obtained from the females fed the purified diet lacking in riboflavin. Ten showed congenital malformations on gross inspection, including shortness of the mandible, tibia, fibula, radius, and ulna, and fusion of ribs with sternal centers of ossification, syndactylism, brachydactylism, and cleft palate.

Subsequently, Warkany and Deuschle[72] performed a histologic analysis of serial sections of dentofacial structures from 15 abnormal and 15 control neonates. Riboflavin deficiency primarily affected the membranous skeleton apparently by interfering with the proper multiplication of mesenchymal cells on or about the 13th day of intrauterine life. The deficiency blocked normal growth and development of many skeletal areas, including the bones of the face where the malformations were expressed as brachygnathia, cleft palate, and growth retardation of the lower incisors.

Deuschle et al.[73] followed the development of the dentofacial complex in rats with prenatal riboflavin deficiency that had appeared almost normal at birth. From the 1st to the 16th day of pregnancy, 46 pregnant Wistar rats were fed a riboflavin-deficient diet. On the 12th, 13th, and 14th day, they were also given 2 mg of the riboflavin antagonist galactoflavin, in 1 mℓ saline by i.p. injection. Except for a slightly twisted tail, 6 of the 219 young born to these dams were normal on inspection at birth. These six animals were followed for 18 months.

Malocclusion of the incisor teeth became apparent at weaning (21 days). The rats were then given a powdered chow diet adequate for growth, but the incisor teeth had to be cut regularly to permit food ingestion and normal growth. The upper and lower jaws became retrognathic at 2 months of age. The maxilla was less affected than the mandible, eventuating into a maxillomandibular disproportion. In all six animals, the upper incisors deviated to the right and the lower incisors to the left, producing a crossbite.

Rats with mild facial anomalies due to maternal riboflavin deficiency could be raised to maturity. Although the jaw anomalies were only subclinical at birth, they became manifest at the time of eruption of the incisor teeth. The incisal abnormality was so severe that survival was possible only by periodic cutting of these teeth. None of the anomalies was self-correcting during the months when the animals were given an adequate diet. Young obtained from matings of these animals with normal rats of the same strain were anatomically normal.

Riboflavin Studies in Man

Sebrell and Butler[74] induced riboflavin deficiency in a group of adult white women in good general condition by dietary means. The 18 institutionalized volunteers were given a special ration prepared in a special kitchen under the constant supervision of a trained dietitian who weighed every item. All ate at one table. The dietary ingredients were relegated to corn meal, cow peas, lard, cane syrup, salt, pepper, loaf bread, white flour, coffee, cod liver oil, and tomato juice. Vitamin analysis showed the ration to be low in nicotinic acid and almost completely free of riboflavin.

Ten of the women developed symptoms similar to pellagra sine pellagra between the 94th and 130th day. There was maceration of the angles of the mouth, reddening of the lips along the line of closure, and thinning and denudation of the mucosa. The word "cheilosis" was coined to describe the morbid condition of the lips. In addition to the lip lesions, the subjects developed a scaly, greasy desquamation in the nasolabial folds, on the alae nasi, in the vestibule of the nose, and — in a few instances — on the ears and eyelids. In all ten women with cheilosis, the lesions did not respond to the administration of nicotinic acid but did disappear following riboflavin therapy. They recurred after riboflavin was discontinued and responded again when therapy was reinstituted. Of the eight women who did not develop cheilosis by the 139th day, four were given a daily preventive dose of riboflavin and four were not. Of the latter, two developed cheilosis on the 191st and 203rd day, and one had a slight maceration at the angles of the mouth on the 200th day.

Niacin Studies in Animals

Afonsky,[75] using the approach described under riboflavin,[68] identified the lingual expressions of experimentally induced niacin deficiency in the dog. All three niacin-deprived dogs developed glossitis represented by diffuse atrophy of the papillae in the anterior and posterior thirds of the dorsum. The lesions spread rapidly and the tongue assumed a bright red color as the deficiency intensified. Initially, there was a blunting of the filiform papillae followed by enlargement of the fungiform papillae which projected above and over the tips of the filiforms. As the atrophy progressed, the filiform papillae underwent extinction while the fungiforms persisted for some time before disappearing and leaving a smooth glazed dorsum. Circumvallate papillae were not affected.

All lesions responded promptly to niacin therapy. Color returned to normal in 24 hr, and papillary regeneration was detectable in 24 to 48 hr. Within 1 week, the dorsum was recovered with papillae. Regrowth started at the midline and spread out to the margins of the tongue, a direction opposite to that of the onset.

Microscopically, the earliest change was a reduction in keratinization of the filiform papillae leading to loss of keratin tips and projections. The papillae became rounded and covered by a layer of parakeratotic cells. Eventually, the stratum granulosum disappeared, and the residual epithelium was composed of proliferating uniform cells that sent out small numerous offshoots. As the atrophy continued, connective tissue cores in the papillae shrank and were obliterated. There were no breaks in the basement membrane nor rarefaction of the underlying connective tissue. Loss of fungiform papillae was due to simple shrinkage in size without degeneration of epithelium or connective tissue. The niacin deficiency glossitis was accompanied by a hemorrhagic marginal gingivitis, reddening, increased sensitivity, and ulceration of the oral mucosa.[76]

Chittendon and Underhill[77] produced a pathologic condition in dogs with symptomology closely resembling that of human pellagra. The dogs were fed a diet of boiled peas, cracker meal, and cotton seed oil. By varying food intake, the disease was induced in from 1 to 8 months. Onset was generally very sudden starting with severe anorexia and apathy. Within a day or two, the mouth developed a characteristic appearance. Buccal and labial mucosa and the margins of the tongue became covered with pustules that underwent extensive necrosis. Ensuing desquamation was accompanied by pronounced fetor oris and heavy salivation. The symptoms were responsive to addition of adequate amounts of meat to the diet.

Goldberger and Wheeler[78] experimentally created a stomatitis in dogs by feeding a diet that was pellagra-producing in man. The disease was clinically and pathologically indistinguishable from the spontaneous condition of dogs known as black tongue or Stuttgart dog epizootic. Initially there were patches of reddening throughout the mouth, followed by development of irregular foci of superficial necrosis. These had a brownish or grayish tint and were covered with a jellylike film or pseudomembrane that was readily removed by wiping with absorbant cotton. As the stomatitis progressed, there was a markedly fetid odor and sialorrhea.

Smith et al.[79] used a modification of the Goldberger-Wheeler diet to produce canine black tongue 96 times in 53 adult male and female dogs of various breeds. In the acute attack, smears and dark field preparations from the mouth lesions showed large numbers of fusospirochetal organisms. Such organisms are usually present only in small numbers around the teeth of normal dogs. The fusospirochetal counts increased enormously during the acute stage of black tongue and disappeared almost completely after dietotherapy only to recur during the next attack.

Two guinea pigs were inoculated in the groin with 1 m*l* and two with 2 m*l* of fresh material obtained from the mouth of a dog with black tongue. After 6 to 10 days, all four animals had groin abscesses containing the inoculated spirochetes, fusiforms, vibrios, and cocci. These organisms were morphologically identical to those present in noma and in Vincent's angina in man.

Denton[80] compared the tissue changes in experimental black tongue of dogs with those in human pellagra. The dogs had served as test animals in the feeding experiments of Goldberger and associates that culminated in the development of black tongue. They were killed with illuminating gas inhalation and carefully necropsied. Tissues were fixed in Zenker's fluid and stained with Giemsa. Formol fixation and hematoxylin and eosin staining were used in some cases.

The most typical and important lesions developed in dogs that were still in a fair state of nutrition. Effects of partial starvation were seen in three animals killed late in the series. In the mouth, the earliest recognizable lesion was reddening of the oral mucosa that usually began in the floor of the mouth, buccal areas, and inner aspect of the lips. Lesions of greater severity showed marked reddened and raised patches of variable size and shape. They were either red or grayish-green due to superficial necrosis of the epithelium and formation of a surface pseudomembrane. Near natural termination, the whole mouth and pharynx were deep red, swollen, and granular with grayish or greenish discolored lesions.

Histologically, the first change was in a narrow zone just beneath the epithelium and superficial to the subpapillary vascular plexus. In the early lesions, marked by reddening of the mucosa, this zone became rarefied and unduly transparent. Loss in density was due to the appearance of large intercellular spaces, widening of the vessels, and diminution of intercellular material. The initial degenerative changes were complicated by proliferation of perivascular cells with lymphocyte infiltration in the intercellular spaces. Degenerative changes in the epithelium were secondary to those in the stroma. Epithelial cells became pale, intercellular spaces were more readily visible, and the surface cells desquamated, leaving either a single layer of epithelial cells or complete epithelial denudation. With the loss of the epithelium, the lesions became simple inflammatory in type covered by a surface pseudomembrane comprised of fibrin, cell debris, and many forms of bacteria. The oral lesions in experimental black tongue in dogs were thus similar to those of human pellagra in gross and microscopic appearance.

In 1938, Elvehjem et al.[81] isolated the factor necessary for the prevention and cure of black tongue in dogs from liver and identified the compound as nicotinic acid amide. That same year Sebrell et al.[82] confirmed that the administration of either sodium nicotinate or nicotinic acid produced rapid recession of the acute symptoms of black tongue in dogs. Shortly thereafter, Heath et al.[83] demonstrated that nicotinic acid was equally effective in the cure of feline pellagra. Each of six cats placed on a nicotinic acid-deficient diet had lost weight and refused food. They appeared weak, sluggish, apathetic and their heads usually hung lower than the rest of their bodies. They made no effort to move when poked and offered no resistance to forcible opening of the mouth or to other physical manipulation. In each, the oral cavity showed a peculiar but typical appearance manifested by an ulcerated scarlet palate and a tongue intensely reddened at the tip and margins. Thick saliva with an extremely foul odor drooled from the mouth. Each cat had a temperature elevation of 3 to 5°F. Within 8 hr after starting administration of 80 to 100 mg nicotinic acid per day, appetite returned, symptoms subsided, oral lesions disappeared, and normal temperature was restored.

Belavady et al.[84] produced nicotinic acid deficiency black tongue in pups fed diets supplemented with leucine in two separate experiments. In each, the young dogs were given a regimen of washed casein, sucrose, cotton seed oil, and salts. Half of the dogs also received daily supplements of leucine. In the first experiment, the dogs were supplied with all of the vitamins except nicotinic acid; in the second, nicotinic acid was included in the vitamin mix.

All of the leucine-supplemented pups in the first study developed signs of black tongue in 31 to 97 days. There was drooling of ropy, blood-stained saliva accompanied by a foul odor. Ulceration and necrosis of the buccal mucosa was extensive, and the entire oropharynx was markedly hyperemic. Smears of the exudate from the oral lesions stained by Fontana's silver method revealed large numbers of fusospirochetal organisms. In the second study, three of five pups given the leucine-supplemented diet developed signs of severe black tongue within 20 days, with the full-fledged picture apparent at the end of 2 months. The other two pups showed deficiency signs after 12 to 15 weeks with periodic exacerbations and remissions.

Histologically, the leucine-supplemented dogs had gross ulceration of the buccal mucosa with active chronic inflammation and necrosis of the epithelium. Focal degenerative epithelial lesions were also present in the tongue. The pathologic changes were indistinguishable from those of dogs that developed black tongue on a diet of jowar or Indian millet.[85] Leucine apparently produced disturbances in the metabolism of nicotinic acid and tryptophane that interfered with the antipellagrenic action of these compounds.

Dreizen et al.[86] determined the histopathology of niacin deficiency stomatitis in marmosets. Although the stomatologic response to experimentally created niacin deficiency had been elucidated in rats, dogs, and rhesus monkeys, each of these studies were done before folic acid and vitamin B_{12} were found to be essential for the proper maturation the epithelium lining of the alimentary tract. To determine whether the integrity of the primate oral mucosa is affected by niacin lack in the presence of adequate amounts of folic acid and vitamin B_{12}, eight adult cotton top marmosets *(Saguinus oedipus)* were made niacin deficient by exclusion of the vitamin from an otherwise nutritionally complete diet. Two marmosets of comparable age and colony status given the identical diet plus 4.9 mg niacin per animal per day served as controls. The study was continued until the test animals became too weak to eat and were killed with metofane. The control animals were sacrificed in the same manner following the death of the last of the experimental group.

All organs removed at necropsy were fixed in 10% buffered formalin. After fixation, they were embedded in paraffin and sectioned for microscopic examination. The formalin-fixed heads were sawed midsagittally, demineralized in formic acid, cut into tooth-containing blocks, and processed for sectioning. Alternate sections were stained with hematoxylin and erythrosin B, Masson's trichrome, and Weigert's iron hematoxylin.

Marmosets made niacin deficient by long-term dietary deprivation (357 to 728 days) developed a syndrome characterized by anorexia, weight loss, weakness, diarrhea, alopecia, enterocolitis, and stomatitis. The alopecia and dermatitis were confined largely to the tail and extremities; the enterocolitis, to all segments of the small and large intestines; the stomatitis, to the tongue and gingiva. Glossitis was reflected by a reddening of the tongue, ulcerations at the tip, margins, and dorsum, and atrophy and loss of the lingual papillae (Figures 4A and 4B).

Microscopically, the relatively smooth dorsal epithelial surface was composed of a basal cell layer, greatly thickened spinous cell layer, and very thin corneum. Rete pegs were widened, and the lamina propria was diffusely infiltrated with chronic inflammatory cells. Surface continuity was disrupted by microbially infected superficial and deep necrotizing ulcers. Both types were marginated by an infiltrate of polymorphonuclear and mononuclear inflammatory cells. Gingivitis was manifested by reddening, swelling, ulceration, and exudation of the interdental papillae capped by a blood-tinged necrotic slough, with extension of the lesions into the attached gingiva. Histologically, there was partial to complete necrosis of the epithelial and connective tissue components of the free gingiva, which often extended into the attached gingiva and coronal aspects of the periodontal ligaments. The necrotic interdental papillae were surfaced by a pyogenous exudate comprised of fibrin, bacteria, pus cells, erythrocytes, and tissue debris. Subjacent surviving soft tissue was diffusely infiltrated with lymphocytes, macrophages, and neutrophils.

Niacin Studies in Man

Spies et al.[87] developed a method wherein oral mucous membrane lesions in pella-

35

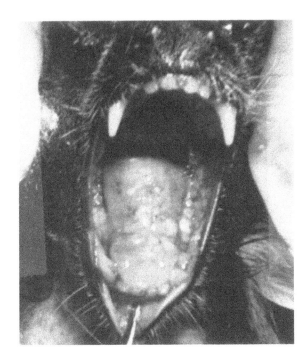

FIGURE 4A.　Ulcerative and atrophic glossitis in a niacin-deficient marmoset. (From Dreizen, S., Levy, B. M., and Bernick, S., *J. Periodontol.*, 48, 452, 1977. With permission.)

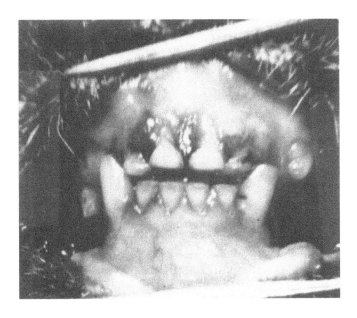

FIGURE 4B.　Ulcerative and necrotizing gingivitis in a niacin-deficient marmoset. (From Dreizen, S., Levy, B. M., and Bernick, S., *J. Periodontol.*, 48, 452, 1977. With permission.)

grins were used as an index for measuring the therapeutic efficacy of various substances. The method involved selection of severely ill pellagrins with classic oral mucous membrane lesions, isolating them and controlling conditions wherein they were limited to a pellagra-producing diet. Those with fiery red mucosal lesions who did not show any improvement on the basal diet served as test subjects. If the substance was of curative value, the lesions healed promptly when adequate doses were added to the basic diet. Swelling and intense redness usually subsided within 48 to 72 hr, often earlier.

The responsiveness of the oral lesions was a much more sensitive marker than that of pellagrous dermatitis. The method was used in 11 pellagrins given nicotinic acid in crystalline form after having been previously used to evaluate liver extract, yeast, and wheat germ against the disease. The group given nicotinic acid by various routes included two endemic pellagrins, three alcoholic pellagrins, and six whose pellagra was secondary to organic disease. Each was selected because of the presence of glossitis, stomatitis, or both. Six had the typical cutaneous lesions; five did not (pellagra sine pellagra). The oral lesions promptly underwent remission in response to nicotinic acid. Pellagrous glossitis, stomatitis, and ptyalism did not reappear while the patients received the vitamin therapy. These studies conclusively demonstrated the specificity of nicotinic acid for the treatment of the oral mucous membrane lesions of pellagra.

Vitamin B₆ (Pyridoxine) Studies in Animals

Levy[88] determined the effects of pyridoxine deficiency on the jaws of mice. In this study, 28 C57 black mice, 28 days of age, were divided into two groups containing equal numbers of males and females. Group I was fed a pyridoxine-deficient synthetic diet; Group II, the same diet containing 54% casein instead of 18%. Animals were kept on the test diets for 1 to 8 weeks before they were sacrificed. Jaws were removed and processed for histologic study. Mice deprived of pyridoxine showed a cessation of mandibular growth, regressive changes in the alveolar bone, and ulceration of the interdental papillae, with replacement of the epithelium by acute and chronic inflammatory cells and necrotic debris. Addition of large amounts of protein to the deficient diet caused the jaw changes to occur earlier and to be more severe.

Afonsky[89] induced pyridoxine deficiency in three dogs by dietary deprivation of the vitamin. All three developed a glossitis during repeated attacks of the disease characterized by a patchy atrophy of the papillae in the anterior two thirds of the dorsum. The ''motheaten'' appearance of the dorsum resembled that of the tongue in riboflavin deficiency. The atrophic areas were irregular in shape and variable in size. Tongue color ranged from pink to red. Bright red was seen in one dog that also developed acute stomatitis and gingivitis.

Microscopically, the degree of papillary change was very variable. Even in the same locus, some papillae showed advanced atrophy while others were almost normal. Papillary atrophy was associated with degenerative changes in the epithelium, notably, hydropic degeneration, loss of keratinization and parakeratosis. Sections of mucosa from animals undergoing a second attack of acute deficiency revealed nerve degeneration and resorption by multinucleated giant cells. The lesions responded well to pyridoxine, with complete regeneration of the papillae and tongue surface in a few days.

Berdjis et al.[90] produced oral and dental lesions in rhesus monkeys made pyridoxine deficient by dietary manipulation. Young, immature male and female rhesus monkeys weighing between 1.8 and 2.5 kg and approximately 12 to 18 months of age were placed on a diet completely lacking in vitamin B₆, but adequate in all other respects. Skulls of animals that died or were killed were fixed in 10% formalin and processed for histology.

All vitamin B_6-deprived monkeys developed acute and chronic gingivitis. In the acute form, the gingiva was swollen, hypertrophic, bright pink to bright red, with a tendency to fill the interdental spaces. In some cases, small superficial ulcerations developed along the margins of the gingiva, floor of the mouth, and palatal mucosa. In the chronic stage, the gingiva was pale, firm, fibrous, grayish red to grayish. Loss of uniform color, partial thickening and partial atrophy of the gingiva, roughness and discoloration of the interdental papillae, with and without ulceration, were noted after long periods of vitamin B_6 deficiency.

Lingual lesions were essentially those of atrophic glossitis. The surface of the tongue was somewhat furrowed in the anterior third and pink to bright red in color. There were a few small erosions in the early stages. Sections of the tongue showed marked atrophy of the epithelium and pale fibrous tissue between atrophied muscle bundles. The jaws of the deficient monkeys showed progressive bone atrophy and atrophy of the soft tissues. Teeth were deformed, shortened, and malaligned. Erosion and decalcification of the teeth were prominent. The pyridoxine deficiency appeared to exert the most deleterious effects during the course of development of the permanent teeth.

Peer et al.[91] found that pyridoxine was partially protective against cleft palate induction in mice by cortisone. Large doses of cortisone administered to pregnant mice just before normal palate fusion in the embryo is often productive of cleft palates in the offspring. To determine the preventive action of pyridoxine, virgin female Swiss albino mice were mated with males of the same strain and subjected to a series of four daily treatments beginning on the 11th day after conception. The mice were given either 0.1 to 0.2 mℓ physiologic saline, 2.5 mg cortisone acetate, 10 μg pyridoxine, or 10 μg folic acid — or combinations thereof — by injection into the thigh muscles. The treated pregnant females were killed 2 days before delivery, the fetuses removed, the cheeks cut, the lower jaw and tongue displaced, and the palate examined under magnification.

Administration of 2.5 mg cortisone daily during the period of treatment produced 109 cleft palates in 127 offspring. There were no cleft palates in those injected with saline. Saline plus cortisone caused 85% of the offspring to have cleft palates. Cleft palate incidence was reduced to 43% when pyridoxine was substituted for saline and to 26% when folic acid was given with cortisone. When both pyridoxine and folic acid were combined with cortisone, there was a further reduction to 15%. Interference by cortisone with the functions of vitamin B_6 as a coenzyme in amino acid metabolism, and with folic acid as a contributor to nucleic acid synthesis, limits the production of protein in the embryo and prevents growth. Supplying these vitamins in sufficient amounts overcomes much of the palatal growth retarding action of cortisone.

Vitamin B_6 (Pyridoxine) Studies in Man

Mueller and Vilter[92] produced pyridoxine deficiency in man by administration of the antimetabolite desoxypyridoxine. Eight patients hospitalized for chronic illness, varying in age from 27 to 87 years, were assigned to two groups. The first received a vitamin B complex-poor diet composed of 3127 cal, 41.7 g protein, 27 g fat, 0.42 mg thiamine, 4.2 mg niacin, 0.6 mg riboflavin, and 0.5 mg pyridoxine plus 60 mg desoxypyridoxine i.m. each day. The second group received the same diet plus 100 mg desoxypyridoxine, later increased to 125 and to 150 mg/day.

All but one of the volunteers developed signs and symptoms ascribable to the experimental regimen. They were given a vitamin mixture containing 25 mg thiamine, 50 mg niacinamide, and 10 mg riboflavin i.m. for 4 to 5 days. Finally, they received 100 to 200 mg pyridoxine i.m., depending on the severity of the clinical condition. One patient was injected with pyridoxine without prior treatment with the other B vitamins. Test diet and desoxypyridoxine were continued throughout the course of the study. Two of the four patients dosed with 60 mg desoxypyridoxine daily developed bilateral angular cheilosis and swelling, redness and soreness of the tongue and buccal mucosa that resembled the stomatitis of niacin deficiency. Lesions developed in 19 to 21 days and disappeared in 48 to 72 hr after administration of 100 mg pyridoxine daily whether B complex vitamins were previously given or not.

Three of the four subjects injected with 100 mg desoxypyridoxine daily developed mouth lesions. One had lesions at the angles of the mouth consisting of excoriation, redness, desquamation of the epithelium, and oiliness by day 15. By day 18, the tongue was swollen and reddened. The stomatitis was alleviated within 48 to 96 hr after the start of pyridoxine therapy. One patient developed ulcers in the buccal mucosa and congestion of the margins of the tongue on day 12. The tongue lesions went on to an acute glossitis characterized by a swollen, diffusely reddened tongue with patchy areas of congested papillae at the margins. The picture resembled pellagrous glossitis, but was pyridoxine responsive. The third developed a unilateral cheilosis on day 31 that responded to pyridoxine by healing in 96 hr. Seborrhea-like lesions developed about the eyes, nose, and mouth after 2 to 3 weeks on the antimetabolite. These lesions, too, responded to administration of pyridoxine, but not to any of the other B vitamins.

Folic Acid Studies in Animals

Afonsky[93] precipitated a recurrent folic acid-deficiency state in dogs by exclusion of the vitamin from an otherwise nutritionally adequate experimental ration. During each attack, the tongue became bright red. Low magnification (5×) disclosed small flat scattered areas of papillary fusion on the dorsum. They were covered by a smooth epithelium, and the surface was level with the tips of the surrounding normal-appearing papillae. During subsequent bouts of the deficiency, the dorsum developed small rough spots composed of irregular masses of papillae that projected slightly above the surface of the neighboring papillae.

Histologically, there was a general tendency for epithelial proliferation on the dorsum associated with papillary atrophy. The outgrowths were especially massive between adjoining papillae and were comprised of epithelial whorls with keratinization. Central areas showed beginning necrosis. There were occasional breaks in the basement membrane and invasion of the epithelium by phagocytic cells. Some of the lingual papillae were fused; others, partially atrophied. In areas of papillary atrophy, the submucosa was thickened and covered with an atrophic epithelium. Regional nerves showed degeneration and resorption by multinucleated giant cells. Although the redness of the tongue subsided during folic acid therapy, the lesions failed to resolve completely.

Stomatitis is a prominent and virtually inescapable accompaniment of the folic acid deficiency states in man and in monkeys. Exfoliative cytology studies of the oral, gastric, and jejunal epithelium from patients with folic acid deficiency have revealed abnormalities that suggest that the vitamin is essential for the proper maturation of rapidly proliferating epithelial cells.

To establish the effect of folic acid deprivation on the integrity of the oral mucosa, Dreizen et al.[94] performed a histopathologic study in marmosets rendered folic acid deficient by dietary and dietochemical means. In this study, 15 healthy adult cotton ear marmosets *(Callithrix jacchus)* were restricted to a folic acid-free purified diet that met all of the other known nutritional needs of the animals. Five of the marmosets also received 0.1 mg methotrexate three times a week by i.m. injection. Three animals given the test diet supplemented with 0.1 mg folic acid per animal per day served as controls. Hematologic changes were monitored from venous blood samples drawn at 3-month intervals and when the test animals became moribund. They were then killed with ether. Controls were sacrificed in like manner after the death of the last of the experimental group. At necropsy, all tissues were fixed in 10% neutral formalin, embedded in paraffin, sectioned, and stained with hematoxylin and erythrosin B, Masson's trichrome, PAS, and Weigert's iron hematoxylin. The fixed skulls were halved midsagittally, cut into tooth-containing blocks, decalcified, and processed in the same manner as the soft tissues.

Ingestion of the folic acid-free diet produced a syndrome characterized by weight loss, lassitude, alopecia, diarrhea, anemia, leukopenia, granulocytopenia, cheilosis, and mucous membrane lesions in the mouth and GI tract that culminated in debilitation, prostration, and death in 59 to 136 days. Methotrexate hastened the onset and progression of the deficiency state. Bone marrow smears of the folate-deficient marmosets showed arrested maturation of the erythroid and myeloid elements. The most noteworthy histopathologic change in the oral mucosa was a ballooning enlargement of the epithelial cells comprising the stratum granulosum and stratum spinosum. The delimiting membrane of the cell was thickened, the cytoplasm was finely dispersed and lightly stained, and the nucleus had a lacy appearance with fragmentation and clumping of chromatin and diminution in staining intensity (Figure 5A).

Keratohyaline granules in the cells of the stratum granulosum were greatly reduced in number and size. Keratin-forming ability was either markedly diminished or completely abolished, evidenced in the tongue by blunting and shortening of the filiform papillae and thinning of the overlying stratum corneum and in the gingival and palatal mucosa by partial or complete absence of cornification.

The deficient animals had shallow ulcers on the dorsum and undersurface of the tongue, lips, pharynx, and gingiva. There were superficial microscopic ulcerations in the GI tract where the intestinal villi were thickened, blunted, widened, swollen, occasionally fused, and always heavily infiltrated with acute and chronic inflammatory cells. Mucosal integrity of the control animals was protected by inclusion of daily supplements of folic acid. The major defect in folic acid deficiency at the cellular level is impaired production of DNA expressed as inefficient mitosis, increase in cellular stroma, and asynchronism between protein synthesis and cell division. Marmosets deprived of folic acid had an exceedingly high incidence of oral infections. These were most severe in the animals that were also given injections of methotrexate. The marked propensity to oral infections was attributed to the breakdown in tissue barriers to infection and to the immunosuppression resulting from deficiency-induced leukopenia (Figure 5B).

Lesions clinically indistinguishable from the cheilosis of ariboflavinosis have been reported in humans deficient in niacin, pyridoxine, pantothenic acid, folic acid, vitamin B_{12}, vitamin C, vitamin A, iron, and protein. The lesions are essentially identical in development and in gross appearance in each of the deficiency states. Although the clinical characteristics of nutritional deficiency cheilosis have been extensively documented in man, histologic examination has not been forthcoming because of cosmetic considerations.

In establishing the nutritional requirements of cotton ear marmosets (C. jacchus), animals restricted to a folic acid-free diet developed a deficiency syndrome in which cheilosis was a prominent feature.[94] The histopathology of these experimentally produced lesions has been detailed by Dreizen and Levy.[95] Severity of the cheilotic lesions ranged from mottling, reddening, swelling, and thinning of the commissures through maceration and denudation of the epithelium to fissure, crust formation, and slough. In the most severe cases, the lesions extended radially onto the surrounding skin for a distance of about 3mm and were accompanied by a generalized cheilitis. In all instances, involvement was bilateral, essentially symmetrical, and progressive in course (Figures 6A, 6B, and 6C).

Histologically, the earliest change was a leukocytic infiltration of the epithelium at the mucocutaneous junction. The infiltrate was considerably more pronounced on the

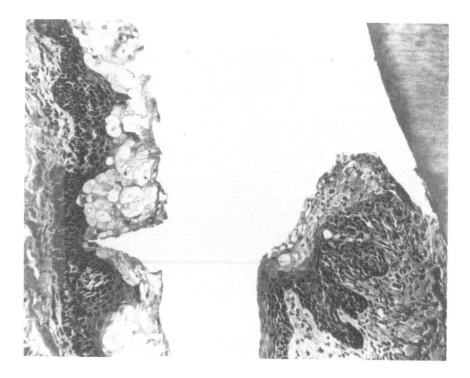

FIGURE 5A.   Ballooning of epithelial cells in buccal (left) and gingival (right) mucosa of folic acid-deficient marmoset. Note absence of keratinization on gingival surface. (Hematoxylin and erythrosin B stain; magnification × 140.) (From Dreizen, S., Levy, B. M., and Bernick, S., *J. Dent. Res.*, 49, 616, 1970. With permission.)

mucosal side of the junction. Cells comprising the stratum spinosum and stratum germinativum had greatly enlarged nuclei. Most of the leukocytes were polymorphonuclear with occasional lymphocytes. Leukocytic infiltration was followed by edema, degeneration, erosion, and loss of the epithelium at both sides of the mucocutaneous border. This stage was marked by sloughing of the epithelium with relatively few inflammatory cells at the base of the developing ulcer. The slough contained numerous lymphocytes, polymorphonuclear leukocytes, and erythrocytes embedded in a fibrinous matrix. In the more advanced lesions, the epithelial continuity was replaced by a crust populated with dense collections of acute and chronic inflammatory cells and ghost epithelial cells. Underlying connective tissue was heavily infiltrated with lymphocytes, polymorphonuclear leukocytes, and plasma cells, the penetration extending to the labial musculature. Regional blood vessels were dilated and engorged. Finally, deep fissures containing necrotic debris, desquamated epithelial cells, inflammatory cells, and bacteria developed at the site of the former mucocutaneous border. The fissures were lined by accumulations of round cells that extended through the depth of the connective tissue into the muscle layers of the labial commissure.

Shklar[96] examined the effects of methotrexate administration on the oral mucosa of hamsters. This study was prompted by clinical observations that patients receiving methotrexate frequently develop oral mucosal lesions. Development and severity of such lesions is apparently related to dose and duration. In this study, 20 male hamsters were given s.c. injections of 0.1 mg methotrexate three times weekly; 20 male hamsters given similarly scheduled injections of physiologic saline were used as controls for the 4-week study. At the dose employed, the only gross change in the oral mucosa was increased reddening. Microscopically, there was epithelial atrophy represented by a

FIGURE 5B. Ulceration and secondary infection of undersurface of tongue in a folic acid-deficient marmoset. (Hematoxylin and erythrosin B stain; magnification × 140.) (From Dreizen, S., Levy, B. M., and Bernick, S., *J. Dent. Res.*, 49, 616, 1970. With permission.)

diminution in the thickness of the stratified squamous epithelium, a flattening of the rete pegs, and a decline in the width of the stratum corneum. Underlying corium showed a notable degeneration of connective tissue collagen. Collagen bundles were clumped and stained deeply compared to the controls.

Nelson et al.[97] found that maternal folic acid deficiency during gestation is productive of multiple congenital anomalies including cleft palate in the offspring. Long-Evans female rats, 3 to 4 months of age, were bred with normal males and placed on a purified folic acid-deficient diet containing succinylsulfathiazole and x-methyl-pteroylglutamic acid at different times during gestation. Groups of animals were started 7 to 15 days after breeding. All rats were necropsied a few hours after delivery or the day before expected parturition (day 21). Fetuses were examined for macroscopic abnormalities, and the uterus was checked for normal or resorbing implantation sites.

Control animals maintained on the identical diet supplemented with a high level of synthetic folic acid (50.5 mg/kg diet) showed no impairment in the reproduction performance and no macroscopic or microscopic abnormalities in the young. Instituting folic acid deficiency in the rat as late as 7 to 9 days after breeding invariably culminated in fetal death through resorption. Delaying the deficiency until 11 days after mating resulted in 95% of the animals littering young with multiple congenital anomalies, notably, skeletal defects like cleft palates and syndactylism, retarded development of viscera — especially the kidneys and lungs, and eye cataracts. Starting the deficiency between 9 and 11 days after conception resulted in approximately 40% of the animals producing litters with a few markedly abnormal young. Delaying the deficiency to 12 or 13 days of gestation yielded 100% litters with the young evidencing mild degrees of edema and visceral retardation, but almost no abnormalities. When folic acid deficiency was not started until 15 days after mating, all the young were normal.

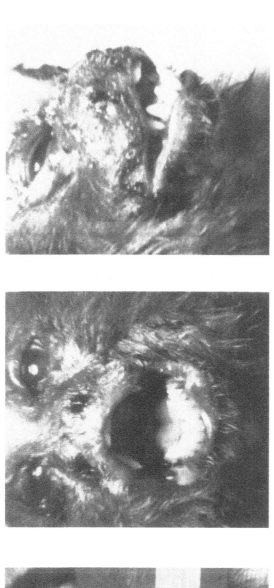

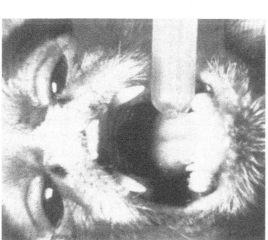

FIGURE 6. (A) Angles of the mouth in a well nourished marmoset, (B) folic acid-deficient marmoset showing moderate cheilosis with desquamation of epithelial cover, (C) folic acid-deficient marmoset with severe cheilosis showing fissuring. (From Dreizen, S. and Levy, B. M., *Arch. Oral Biol.*, 14, 577, 1969. With permission.)

Folic Acid Studies in Man

Vogel and Deasy[98] studied the preventive effect of folic acid on experimentally produced gingivitis in 16 volunteer dental students. The students were given a thorough dental prophylaxis and placed on a strict home care regimen to remove dental plaque. They were randomized into two groups of eight each 10 days later. The test group was given 2 mg of crystalline folic acid daily; the control group received a placebo. Gingival inflammation and plaque accumulation were assessed and the students instructed to stop using oral hygiene procedures for one half of the mouth while continuing care for the other half. On day 14 they were reassessed, and hygiene was reinstituted for 14 days when a third examination was made.

Gingival exudate levels and gingival bleeding indexes on day 0 indicated that all participants had attained a high level of gingival health. On day 14 the control group had a significantly higher degree of gingival inflammation than the test group. Both had about the same amount of plaque and higher gingival bleeding indexes than at the start. A significant degree of increase in gingival exudate flow was found in the controls. On day 28, there were no meaningful differences between groups for any of the parameters measured. After home care was reapplied, appreciable diminutions in gingival inflammation, gingival bleeding, and plaque levels were noted. The findings indicate that a pharmacologic dose of folic acid may increase gingival resistance to local agents productive of inflammation.

Rose[99] determined serum folate levels in 50 patients with angular cheilosis and in 47 healthy controls. Age and sex distribution were similar in both groups. Mean serum folate concentration was significantly lower in the cheilosis group. In addition, 18% of the cheilosis patients had low serum folate levels while the values for each of the controls was within normal limits. In two patients with low serum folate, treatment with folic acid caused a disappearance of the lesions within 1 month, suggesting a causal relationship.

Pantothenic Acid Studies in Animals

Afonsky[100] created a pantothenic acid deficiency in two dogs by dietary deprivation, expressed clinically by similar lingual lesions. In each, the dorsum of the tongue was light gray with an atrophy of the filiform papillae. Fungiform papillae stood out as prominent bulbs. Papillary changes were confined to the anterior part of the dorsum. A few sharply demarcated areas of complete atrophy ranging from 3 to 4 mm in diameter were seen during a second attack of the deficiency.

Microscopically, the filiform papillae in the anterior part of the tongue were small but regular in shape. The epithelium was thin, poorly differentiated, and lacked the usual stratification of cell layers. Basal cell layer was normal, but the stratum spinosum was almost entirely absent. The cells varied in size and evidenced hydropic degeneration. Nuclei were irregular in size and shape. Stratum granulosum was thick in some places but contained only a few keratohyaline granules. There was a complete loss of keratinization with the surface covered by a thick layer of parakeratotic cells. Occasionally, a piling up of cells caused disfigurement of the papillae. Nerves were necrotic and in the process of resorption by multinuclear giant cells. Response to specific vitamin therapy was prompt; normal color and papillation returned within 1 week.

Wainwright and Nelson[101] detailed the changes in the oral mucosa that accompanied experimental pantothenic acid deficiency in young rats. In this study 17 rats were placed on a pantothenic acid-deficient diet at birth; 9 were started at weaning on day 21. Five of the nine were also given a high amount of fat in the diet in place of carbohydrate. All test rats were sacrificed just before imminent death. Control animals receiving the same purified diet fortified with pantothenic acid, and normal rats fed natural foods were sacrificed at the same ages as the test animals. All were necropsied and the oral tissues processed for histologic examination.

Rats deficient in pantothenic acid from birth showed hyperkeratosis of the enamel and mouth epithelium at 21 to 48 days of age. Necrosis of the epithelium began at 48 days and advanced deeper and deeper with increasing age. In rats surviving for more than 72 days, the necrotic process destroyed all epithelial cells in the interdental papillae and underlying connective tissue to the crest of the alveolar bone. Resorption of the alveolar crest occurred in a few rats. At each stage up to complete destruction of the epithelium, there was a proliferation of the basal cells, presumably, in an attempt to protect the underlying tissue from exposure to the oral cavity. Epithelial necrosis was not accompanied by an inflammatory reaction regardless of the degree of destruction. Rats given the pantothenic acid-deficient diet from weaning showed the same epithelial necrosis as animals deficient from birth. These changes developed on the basal diets whether high in fat or in carbohydrate.

Ziskin et al.[102] investigated the oral reactions in rats fed a synthetic diet complete in every respect but containing suboptimal amounts of pantothenic acid with or without 0.08% zinc as zinc carbonate. The study series was comprised of 75 female rats. Of the group, 47 were restricted to a pantothenic acid-deficient diet plus zinc carbonate; 10 were given only the pantothenic acid-deficient diet; 18 served as litter mate controls. Animals judged to be in a terminal state were sacrificed and necropsied. The tongues were removed and the heads split along the midline. Half the head was X-rayed, and half was prepared for histologic examination.

Rats on the suboptimal pantothenic acid diet were depressed in weight. The added zinc caused signs of severe poisoning: alopecia, dacryohemmorrhea, rusty, scrubby, and oily hair, anemia, and granulocytopenia. Optimal amounts of pantothenic acid were completely protective against the zinc-induced skin lesions. Ulceration of the tongue, buccal mucosa, and palate occurred in about 50% of the test animals. The tongue ulcers were mainly on the dorsum close to the midline between the tip and the posterior third and occasionally on the undersurface. Crusting and ulcer formation at the angles of the mouth was evidenced by about 67% of the animals. The ulcers were round or semicircular, frequently surrounded by a raised border. There was a close correlation between the occurrence of mouth ulcers and presence of severe skin lesions. Mouth ulcers began to appear in the 5th to 7th week of the experiment. About 51% of the rats had hyperkeratosis and loss of the filiform papillae in the tongue.

Many of the experimental animals had decolorization of the incisors and defective incisal dentin formation with globular calcification, wide noncalcified zones, and irregular outline of the dentinal border. Except for occasional poor dentinal calcification, there were no other alterations in the molar teeth. About 60% of the rats had evidence of periodontal involvement either on X-ray or microscopy, with no relation to either age or severity of skin lesions.

Levy[103] explored the effects of pantothenic acid deficiency on the mandibular joint and periodontium of mice. In this study, 20 male C57 black mice, aged 25 to 28 days, were relegated to a synthetic diet lacking in pantothenic acid for 1 to 4 weeks. Nine males of the same age were given the test diet for 4 weeks that was then replaced by Purina® laboratory chow for varying periods. Additional mice were fed either the complete synthetic diet or laboratory chow.

All animals on the deficient diet failed to gain or actually lost weight. Ventral and facial alopecia developed between weeks 3 and 4 and graying of the hair after week 4. Microscopic examination of sections prepared from the jaws removed at necropsy revealed that pantothenic acid deficiency impaired proliferation and maturation of cartilage cells and arrested osteogenesis in the growth plate of the mandibular condyle. Condylar changes were accompanied by resorption of the alveolar bone, narrowing of the interdental bony septae, widening of the periodontal ligaments, and lowering of the alveolar bony crests. There was epithelial downgrowth along the roots of the teeth and an increase in size of the epithelial rests.

Frandsen et al.[104] performed a similar experiment in 20 Long-Evans rats maintained on a pantothenic acid-deficient diet from birth. Histologic changes in this series were characterized by marked interference with chondrogenesis and osteogenesis. The late stages of the deficiency were dominated by necrosis of the articular capsule and articular disk, together with fibrous tissue proliferation into the glenoid fossa productive of destruction of the temperomandibular joint. Severity was related more to individual response than to duration of deficiency. Bone resorption was very extensive, leading to almost a complete absence of trabeculae in the head of the condyle and to partial disappearance of the lamina compacta in the ramus. Concurrently, there was osteophytic growth from the periosteal surface of the ramus and squamosal bone. All changes were prevented by the administration of calcium pantothenate.

Nelson et al.[105] described the teratogenic effects of pantothenic acid deficiency in rats. Female Long-Evans rats, 60 to 65 days of age, were placed on a pantothenic acid-deficient diet from the day of breeding. Others were mated after approximately 4 to 20 days on the deficient diet and continued thereon. Control animals received the test diet for 18 to 20 days before breeding, then supplemented with pantothenic acid throughout gestation. Each group contained 20 to 24 rats.

To determine the effects of pantothenic acid deficiency during the first half of pregnancy, rats of the same age were fed the deficient diet for the first 12 to 14 days of gestation and then given the complete diet. To test the effects of omega-methyl-pantothenic acid, rats were maintained on the deficient diet for the first 9 to 11 days, then fed the antimetabolite and test diet for 2 or 3 days, followed by complete diet for the remainder of gestation. All rats were necropsied on the 21st day of gestation, 1 day before parturition. Fetuses were removed by section and examined microscopically for abnormalities.

Multiple congenital defects were produced when the deficiency was instituted on the 1st day of gestation or 4 to 10 days before breeding, and continued throughout pregnancy. Limitation of the deficiency to the first 12 to 14 days of gestation resulted in few anomalies. Addition of the antimetabolite to the deficient diet for 2 or 3 days accentuated the deficiency and produced more fetal deaths and anomalies, including cerebral, skin, eye, aortal, and heart defects, digital hemorrhages, edema, hydronephrosis, hydroureter, club foot, and cleft palate.

Biotin Studies in Animals

Ziskin et al.[106] delineated the oral manifestations of biotin deficiency in rats fed a synthetic diet devoid of biotin in which egg whites replaced casein as the protein source. The diet was fed for 12 weeks. Surviving animals were then sacrificed and the oral structures examined. The major sign of biotin lack was the appearance of clear and hemorrhagic subepithelial vesicles on the dorsum of the tongue that were not present in the control group.

Collins et al.[107] studied the changes in the temperomandibular condyle of Long-Evans rats made biotin deficient by dietary deletion. The deficiency produced little change in the chondrogenic process, but greatly reduced appositional growth around the trabeculae. There was also an accompanying replacement of the bone marrow by fibrous tissues.

Biotin Studies in Man

Sydenstricker et al.[108] fed four volunteers (three white men and one black woman) a diet designed to contain a minimal amount of biotin. The ingredients were 125 g rice, 80 g patent white flour, 75 g farina, 250 g cane sugar, 32 g lard, 10 g washed butter, and 25 g lean beef. To these was added 200 g of dehydrated egg white. Total caloric content was 2888 per day. Egg white was given in solution with each meal and usually furnished in excess of 30% of the daily caloric intake. Since the diet was poor in B complex vitamins, supplements of thiamine, riboflavin, nicotinic acid, pyridoxine, calcium pantothenate, vitamin A, ferrous sulphate, and calcium lactate were given daily.

All volunteers were in good condition and free from signs and symptoms of avitaminosis when the experiment began. During the 3rd and 4th week, they developed a fine, scaly, nonpruritic dermatitis that disappeared spontaneously. During the 7th and 8th weeks all showed a striking grayish pallor of the skin out of proportion to the blood picture interpreted as evidence of peripheral vasoconstriction. During the same period, the three white subjects showed definite atrophy of the lingual papillae; patchy and productive of a "geographic" tongue in the first, generalized in the second, and confined to the lateral third of the dorsum in the third. The black subject showed no changes in the tongue until the 14th week when a rather rapid denudation of the lingual papillae began. In each, the tongue remained pale throughout without any of the capillary engorgement characteristic of pellagra or ariboflavinosis. Other evidence of biotin deficiency included anorexia, nausea, vomiting, mental depression, muscle pains, paresthesias, and precordial discomfort. All signs and symptoms cleared rapidly following parenteral administration of a concentrate representing 150 to 200 μg biotin per day.

Baugh et al.[109] documented a diet-induced biotin deficiency in a 62-year-old white female. She had consumed six raw eggs and 2 qt of skim milk daily for 18 months. This diet had been recommended for the treatment of Laennec's cirrhosis. After a few weeks, the subject developed anorexia, dysphagia, and soreness of the tongue and lips. She continued to eat the eggs and milk and took a high potency multivitamin capsule daily that was devoid of biotin. She also consumed several brewer's yeast tablets per day that provided an unknown amount of biotin and received 100 μg B_{12} monthly by injection.

Due to the avidin content of the egg white, the stage was set for the development of an uncomplicated biotin deficiency. Avidin is a protein capable of binding with biotin rendering it nutritionally inactive. Although the patient had noticed mild dryness of the skin and occasional soreness of the mouth and lips, it was not until several months prior to the hospital admission that the symptoms increased in severity. The tongue became sore and reddened and the lips were fissured, encrusted, and bled occasionally. She had mild dysphagia, nausea, and vomiting. Associated with these symptoms were increasing lassitude, malaise, exertional dyspnea, easy fatigability, and substernal discomfort partially relieved by belching.

On physical examination, the skin was dry and covered with fine scales; the lips dry, cracked, and scabbed over; the oral mucosa hyperemic; and the tongue magenta colored. Fungiform papillae stood out as prominent reddened knobs. Other physical abnormalities included hepatomegaly, splenomegaly, and mild pitting ankle edema. Intramuscular biotin therapy at a level of 200 μg per day produced a complete resolution of the stomatitis within 4 days.

Vitamin B_{12} Studies in Man

Farrant[110] followed the nuclear changes in the squamous cells of the buccal mucosa in patients with pernicious anemia before and after treatment with vitamin B_{12}. Films of the mucosa were obtained by scraping the inside of the right cheek with a microscope slide. The sample was spread, air dried, fixed and stained with May-Grunwald-Giemsa. The microscopic image was magnified onto a ground glass screen (magnification × 1000). Measurements of the long and short axes were made of cells in the center of the field. For each axis, the means of 100 measurements were used for comparison.

The study group consisted of 25 patients with pernicious anemia examined before and after treatment with vitamin B_{12}. Prior to treatment, the squamous cell nuclei were abnormally large and had irregular asymmetrical outlines. Subsequent to treatment there was a highly significant decrease in nuclear size; the decrease was greatest in the short axis. Not only did the nuclei become smaller, but they also had a more uniform shape. The change to normal size occurred within a few days after treatment was started and preceded the repapillation of the dorsum of the tongue, which is often a prominent feature of the disease.

Stone and Spies[111] demonstrated that vitamin B_{12} is curative for the severe oral mucous membrane lesions that develop in some patients with pernicious anemia. These are usually relieved by liver extract, but not by folic acid. The lesions are characterized by a fiery red excrutiatingly painful denudative glossitis and/or stomatitis. They recurred in the same part of the mouth during each relapse and underwent complete remission with each course of parenteral vitamin B_{12} therapy.

Vitamin B Complex Studies in Animals

Gorlin and Levy[112] determined the changes and course of repair in the mandibular joint and periodontium of vitamin B complex-deficient rats. In this study, 45 Wisconsin strain rats (24 males, 31 females) 19 to 22 days old were distributed into three comparable groups. Group I was comprised of 12 rats that received a stock diet and were killed after 18 to 40 days on the ration. Group II contained 21 animals fed a vitamin B complex-deficient diet for the same period unless poor condition necessitated early sacrifice. Group III consisted of 22 rats given the vitamin B-free diet for 3 weeks followed by stock diet for 5 to 30 days. Test animals and their controls were killed with chloroform at the end of each timed experiment. Heads were disarticulated, fixed in 10% formalin, decalcified, paraffin-embedded, serially sectioned, and stained with hematoxylin and eosin.

The vitamin B complex-deficient animals evidenced diminution and cessation of chondrogenesis and osteogenesis, atrophy of connective tissue, serous atrophy of bone marrow, and resorption of cementum and dentin. Refeeding with stock diet resulted in rapid resumption of chondrogenesis and osteogenesis, remodeling of bone, hyperplasia of fibrous connective tissue, and repair of cementum or bonelike substance.

Chapman and Harris[113] delineated the stomatologic responses in rhesus monkeys to diets deficient in all or part of the vitamin B complex. Animals on the test diets developed gingivitis and an increase in fusospirochetal flora. Some had extensive gingival lesions and ulceration of the buccal and labial mucosa. Oral lesions were first observed on the 57th experimental day. Monkeys maintained on an adequate stock diet did not exhibit oral lesions or an increase in fusospirochetes even when artificial implantation of these organisms was attempted.

Vitamin C Studies in Animals

Glickman[114] identified the effects of acute vitamin C deficiency on the periodontal tissues of the guinea pig. In this study, 16 of 25 young adult guinea pigs weighing from 200 to 300 g were placed on a vitamin C-free diet; the other 9, on the same diet supplemented with 1 mg ascorbic acid per 100 g body weight per day. All were sacrificed after 35 days, and the jaws prepared for histologic study.

Acute vitamin C deficiency produced disturbances in the periodontal membrane and alveolar bone in the absence of gingival inflammation and pocket formation, that resulted in a loss of the supporting tissues of the teeth. The disturbances were reflected by a generalized periodontoclasia or diffuse alveolar atrophy. Deficiency-related alterations in the connective tissue of the marginal gingiva were manifested by edema, collagen degeneration, and hemorrhages.

Subsequently, Glickman[115] elicited the effect of acute vitamin C deficiency on the response of the periodontium to artificially induced inflammation. Inflammation was generated by application of a 10% silver nitrate solution to the gingival sulcus for 30 sec in vitamin C deficient and control guinea pigs. Although comparable pocket formation and localized necrosis occurred in both groups, the associated destructive changes in the underlying tissues were more severe in the scorbutic animals.

Acute vitamin C deficiency alterations in the periodontal tissues accelerated the distributive effects of inflammation. There was a lowering of the gingival resistance to the destructive action of inflammation in the vitamin C-deficient animals due to an absence of a well-formed barrier of collagen fibrils, fibrin and inflammatory cells between the artifically induced injury and the underlying bone. Additionally, the vitamin C-deficient animals demonstrated collagen degeneration, failure to form new collagen, osteoporosis of the alveolar bone, inability to produce a proper bone matrix, and inhibition of fibroblast formation and differentiation into osteoblasts.

Boyle et al.[116] produced diffuse alveolar bone atrophy in guinea pigs by diets deficient in vitamin C that was histopathologically similar to that in the jaws of scorbutic infants. Microscopic examination of sections from scorbutic animals killed while moribund disclosed marked rarefaction of the alveolar bone and widening of the periodontal membrane. The changes were attributed to a failure of the osteoblasts to form bone matrix and of fibroblasts to form collagen.

Dreizen et al.[117] examined the influence of vitamin C deficiency on the primate periodontium. Six healthy adult cotton ear marmosets *(Callithrix jacchus)* were restricted to an ascorbic acid-free purified diet containing all of the other nutritional essentials. Three marmosets received the identical diet supplemented with 25 mg ascorbic acid per animal per day. The test animals were kept on the deficient diet until they became moribund, at which time they were anesthetized, exsanguinated by cardiac puncture, and necropsied. Control animals were killed in the same manner following the death of the last of the test group.

The vitamin C-deprived marmosets developed petechiae and ecchymoses in the skin of the extremities, face, and abdomen after 2 to 4 months on the scorbutic diet. They later evidenced anorexia, lassitude, sustained weight loss, swelling of the eyelids and joints, muscle tenderness, and restricted locomotion. At necropsy, the most striking findings were hemorrhages at the costochondral junctions and in the vertebrae and joint enlargement. Extensive bleeding was also found in the muscles of the extremities and under the periosteum of the ribs and long bones. Microscopically, the hemorrhages were poorly organized.

The teeth of the scorbutic animals were markedly mobile. Interseptal and ligamental blood vessels were congested, and perivascular and perineural spaces were expanded. Periodontal fiber bundles surrounding these spaces were greatly diminished in thickness. With increasing duration of the scorbutic state, the periodontal ligaments became infiltrated with multiple hemorrhages apical to the transeptal fibers. There was continuing enlargement of areas of decreased ligamental density. Fatty bone marrow was replaced by actively proliferating red marrow that in rare instances crowded through the perforations in the cribriform plate to enter the periodontal ligament space. Ultimately, the compact character of the periodontal ligament was replaced by a loosely arranged pattern of thin fibers and fibrils. The periodontal ligament was widened by osteoclastic resorption of the surrounding bone, edema, and diminution in number of attachment fibers. The striking changes in the periodontal ligament were accompanied by relatively minor alterations in the osseous and gingival components of the periodontium (Figure 7).

Hunt and Paynter[118] used a basal semisynthetic diet of known composition, pure l-ascorbic acid supplements, and pair-feeding methods that factored out pathologic al-

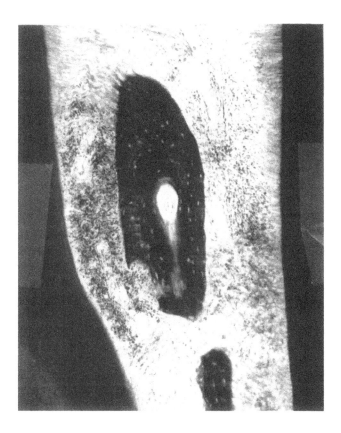

FIGURE 7. Multiple hemorrhages and diminution of collagen fiber density in periodontal ligament of a scorbutic marmoset. (From Dreizen, S., Levy, B. M., and Bernick, S., *J. Periodontol. Res.*, 4, 274, 1969. With permission.)

terations from other dietary deficiencies, to elucidate the effects of ascorbic acid deficiency on the teeth and periodontal tissues of guinea pigs. In this study, 5-week-old animals weighing between 190 and 270 g were divided into four main groups and treated as follows:

Group number	Number of animals	Ascorbic acid mg/day/animal	Experimental period (days)	Type of feeding
I	5	0.0	24	*Ad libitum*
IIa	6	0.4	48	*Ad libitum*
IIb	6	5.0	48	Pair-fed to IIa
IIc	6	5.0	48	*Ad libitum*
IIIa	7	0.4	75	*Ad libitum*
IIIb	7	5.0	75	Pair-fed to IIIa
IIIc	7	5.0	75	*Ad libitum*
IV	5	0.4	48	
		then		*Ad libitum*
		5.0	27	

At the termination of each experiment, the guinea pigs were killed by decapitation and the heads were fixed in Bouin's solution. The tissues were demineralized, embedded in paraffin, sectioned, and stained with hematoxylin and eosin, toluidin blue, van Gieson and silver impregnation.

All animals fed less than 5 mg ascorbic acid per day developed paralysis of the hind legs that was directly related to the reduction in dietary ascorbic acid. None of the animals given 5 mg ascorbic acid per day, whether *ad libitum* or pair-fed, had microscopic alterations in any of the tissues. In those given 0.4 mg ascorbic acid, there were notable changes in dentinal structure and mild changes in the pattern of alveolar bone deposition and resorption. The latter were seen only after 75 days of feeding. Tissue changes caused by limitation to 0.4 mg ascorbic acid per day for 48 days were reversed by supplements of 5 mg/day for 27 days. Animals restricted to the ascorbic acid-free diet showed the most severe disruptions. Hemorrhages were present in the pulps and periodontal ligaments. Collagen, cementum, dentin, and bone formation were impeded. Bone resorption was excessive with the severity apparently pressure related. Enamel matrix failed to form where predentin formation was defective.

The histopathologic effects of acute and chronic vitamin C deficiency on the teeth of guinea pigs were detailed by Boyle et al.[119] Animals were kept on a vitamin C-free diet plus 0.3 to 2 mg vitamin C daily. Negative controls on basal diet alone and positive controls on larger amounts of ascorbic acid and greens were included in the experimental design. In acute scurvy (basal diet alone), there was marked atrophy of the cells of the dental pulp. Odontoblasts became morphologically indistinguishable from other pulp cells. Dentin deposition ceased. On 0.3 mg ascorbic acid daily, tooth formation continued normally; mantle dentin deposited at basal ends of the incisors was normal with no discernible change in the regional odontoblasts. Farther incisally, the odontoblasts assumed a papillary arrangement about the capillaries in the pulp. Farther apically, long spicules of dentin projected into the pulp cavity.

There were hemorrhages in the enamel organs and periodontal membranes, with occasional enamel hypoplasia. Tooth structure formed during the experimental period was extremely fragile after eruption into the oral cavity. Similar but less marked changes occurred in animals given 0.5 and 1 mg ascorbic acid daily. No pathology was found in those receiving 2 mg ascorbic acid per day.

Kalnins[120] investigated developmental disturbances in the enamel in scurvy in an attempt to determine whether they were due to a direct effect of the deficiency on amelogenesis or were secondary to scorbutogenic deficiencies in dentin formation. Changes in the enamel in prescurvy and in clinical scurvy were studied in 62 guinea pigs that had received a scorbutogenic diet. Enamel alteration during the production of scurvy, subscurvy, and first stage of cure of scurvy were studied in 110 guinea pigs given 0.1, 3, or 25 mg ascorbic acid per animal per day for 1 to 20 days following a scorbutic duration of 13 to 23 days. The jaws were removed at necropsy, embedded in paraffin or celloidin, sectioned, and stained with hematoxylin and eosin.

The most common scorbutic changes in the enamel consisted of (1) aplasia and hypoplasia in the molars resulting from ameloblast dysfunction caused by a reduction and complete cessation of dentin development, (2) hypoplastic formation of the enamel matrix in the molars and incisors due to detachment, pressure atrophy, or distortion of ameloblasts during growth, and (3) partial or complete disturbance of maturation of enamel in the molars and incisors caused by occlusal trauma or bending of the tooth that affected the ameloblasts. Reduction and cessation of dentinal growth was found to be the primary and typical cause of scorbutic changes in the enamel. Differentiation and secretory activity of the ameloblast is not disturbed by scurvy. Only the capacity to produce enamel matrix is lost because of the absence of dentin.

Avery et al.[121] determined the modifications in the fine structure of the guinea pig incisal pulp during experimental scurvy. In this study, 20 guinea pigs weighing about 250 g were pair-fed a scorbutogenic diet for 30 to 35 days. Control animals received daily oral supplements of 20 mg ascorbic acid. At the end of the test period, each animal was anesthetized with ether and the mandible rapidly dissected out and bisected at the midline. The entire incisor was exposed (including the growth end), removed from the bone, and dipped briefly in fixative. The partially calcified dentin at the proximal portion of the tooth was removed and the pulp extracted from the hard shell dentin by a gentle pull with a forceps. Small pieces of pulp were cut and fixed for electron microscope examination.

In the normal animals, the differentiating fibroblasts in the pulp were stellate, with long cytoplasmic processes. They had a well-developed, rough-surfaced endoplasmic reticulum, Golgi complex, and mitochondria. Ground cytoplasm was dense and contained varying amounts of intracellular fibrils measuring approximately 60 Å in diameter. Along the inner surface of the plasma membrane were several vesicles and vacuoles. Extracellular fine fibrils 100 to 120 Å in diameter were found usually in groups surrounded by aggregations of finely granular ground substance. Collagen fibrils 400 to 700 Å in diameter with characteristic cross banding were also located near the cells.

In the scorbutic guinea pigs, the fibroblasts appeared smaller and the intracellular organelles were diminished in number. Many cells were devoid of endoplasmic reticulum which, when present, appeared dilated with a decreased number of ribosomes on the surface. A significant increase in the number of intracellular and intercellular fine fibers was found. Intercellular fibrils were situated near the plasma membrane; some appeared in isolated bundles.

Levy and Gorlin[122] were the first to establish the serial changes in the mandibular condyle during the development of experimental scurvy. In this study, 12 young guinea pigs weighing between 200 and 300 g were placed on a vitamin C-free diet and killed after 2, 3, and 4 weeks. Some received injections of ascorbic acid after 3 weeks and were killed 3 days later. At sacrifice, the heads were removed, formalin fixed, decalcified, blocked, and sectioned.

After the first 2 weeks on the deficient diet, there was an increase in the number of osteoclasts in the zone of erosion. Numerous fibroblastlike cells replaced the hematopoeitic cells in the subchondral area. By 3 weeks, osteogenesis was completely arrested. Calcified cartilage was separated from the marrow by an acidophilic structureless material. The cartilage band was irregular and displayed numerous fractures of the matrix material that lay in disarray in the immediate subchondral area.

Bone marrow in the subchondral region and in the ramus was largely replaced by connective tissue cells. The marrow spaces contained numerous broken bone spicules surrounded by focal accumulations of osteoclasts. Lateral width of the ramus was reduced. An acidophilic amorphous material was present on both sides of the miniscus. Muscles and periosteum were edematous, and there were small hemorrhages between the muscle bundles and below the periosteum.

After 4 weeks on the scorbutogenic diet, all changes were further accentuated. Frank hemorrhages were present in the temperomandibular joint. There was a reduction of the edema 3 days after ascorbic acid therapy, as well as a tremendous growth of osteophytic bone subperiosteally. Many of the osteophytes resembled cartilage. The pink staining material in the subchondral area gradually disappeared, and the fractures at the cartilage-shaft junction were undergoing repair, as evidenced by callus formation.

Irving and Durkin[123] compared the changes in the mandibular condyle with those in the proximal tibial epiphysis during the onset and healing of experimentally produced scurvy in guinea pigs. In this 30-day study, 104 animals of both sexes weighing approximately 300 g were used. Animals were killed at intervals beginning on day 16. Of the group, 28 control animals were given 30 mg ascorbic acid daily by mouth; 39 animals were maintained on the vitamin-deficient diet throughout the experiment; 37 animals were given the scorbutogenic diet and a s.c. injection of 30 mg ascorbic acid from 3 to 30 days before sacrifice. The tibia and temperomandibular joint from one side were fixed in formol saline, embedded in paraffin, sectioned, and stained with hematoxylin and eosin; the opposite members were similarly prepared, embedded in gelatin, sectioned, and stained with Sudan black.

Classic scorbutic changes, notably, decreased differentiation of the cells of the epiphyseal cartilage, arrested osteogenesis, and accumulation of fibroblastlike cells productive of a large gerustmark were seen in the tibia by the 22nd day. The sequence of events in the mandible was entirely different. No gerustmark was formed, and the cartilage became gradually wider as osteoclastic resorption slowed and stopped. On giving ascorbic acid, osteoclasts reappeared on the face of the wide condylar cartilage within 24 hr, and by 72 hr the cartilage had become virtually normal in width. The changes in the condyle during scurvy and healing are thus more consistent with a remodeling role than with a growth center function.

In a follow-up study, Durkin et al.[124] compared the circulatory and calcification changes in the mandibular condyle and tibial epiphyseal cartilages of guinea pigs under similar experimental conditions. In the scorbutic tibial epiphyseal cartilage, the vascular pattern was one of progressive disorganization and disruption, with an overall reduction in both metaphyseal and epiphyseal blood supply to the cartilaginous plate. In the condylar cartilage, the onset of scurvy caused a similar but less drastic disruption of capillary distribution.

Mineralization of the scorbutic tibial epiphyseal cartilage included the entire hypertrophic zone which was not the case in the scorbutic condylar cartilage. In the tibias of the ascorbic acid-treated animals there was a gradual recovery of the normal capillary and calcification patterns. The reconstruction process was completed by 72 hr postinjection. Condylar cartilage showed a similar but faster repair of the circulatory pattern coincident with a return to normal calcification. The absence of a gerustmark and trummerfeld zone beneath the condylar cartilage may be responsible for the less dramatic changes and faster repair rate in this area when compared to the tibial epiphysis.

Turesky and Glickman[125] made a histochemical evaluation of gingival healing in experimental animals on adequate and vitamin C-deficient diets. In this study, 44 guinea pigs weighing 224 to 362 g were assigned to three protocols: Group I, complete diet and gingivectomy; Group II, vitamin C-deficient diet; Group III, vitamin C-deficient diet and gingivectomy. Animals in Groups II and III showed clinical evidence of scurvy by day 14. Gingival biopsies were obtained from Group II after 18 and 24 days on the scorbutogenic diet. Gingivectomies were performed on the Group III animals after 14 study days. Specimens of healing gingiva were removed between 4 and 21 days after gingivectomy. The same procedure and schedule were followed for the Group I control animals. The diet of Group III was supplemented with 10 mg ascorbic acid per day for 10 days following the postoperative experimental period. Gingival tissues were either fixed in chilled absolute alcohol for 48 hr at 5°C or in alcoholic formalin for 24 hr. Serial sections were stained for ground substance, basement membrane, and glycogen.

The effect of acute vitamin C deficiency on the unoperated gingiva was confined to the connective tissue and consisted of a reduction in ground substance and glycogen with beading and fragmentation of the collagen fibers. In the control animals, ground substance first appeared in the healing gingiva on day 4 as a thin amorphous film overlying thin collagen fibers. Later, the ground substance was denser and the collagen fibers coarser. In the scorbutic animals there was no formation of ground substance during the healing period. Instead, there were gel-like pools of procollagen. The basement membrane of the healing gingiva in the controls was broad and ill-defined in the early stages of healing but became thinner, more definitive, and stained less intensely as healing progressed. In the scorbutic animals, the basement membrane of the healing gingiva remained broad, ill-defined, and stained deep pink throughout the study period. In both groups, the initial extension of the epithelial spurs over the wound area had no basement membrane. Appearance of the basement membrane was associated with maturation of the underlying connective tissue.

Since pharmacodynamic doses of vitamin C act as detoxicants, Kalnins[126] studied the role of large doses of ascorbic acid in the prevention of bone and tooth lesions resulting from X-irradiation. Two groups of guinea pigs were fed a scorbutogenic diet plus one or two drops of fish oil twice a week. One group received a supplement of 1 mg ascorbic acid s.c. every other day. This protected the animals from the symptoms of frank scurvy, but was not sufficient for maintaining normal rate of deposition and normal structure of dentin. Ascorbic acid supplements were started 20 days before irradiation. The second group received 50 mg ascorbic acid s.c. every other day, a dose that provided full protection against scurvy. Hypovitaminotic and control guinea pigs were given either a single dose of 800 R or 900 R, or fractionated doses of 1000 R, 1200 R, 2600 R, and 4200 R X-irradiation over a 14- to 29- day period. The animals were then killed, the mandibles roentgenographed, fixed in 10% formalin, decalcified, paraffin-embedded, sectioned, and stained with hematoxylin and eosin.

At exposures of 1200, 2600, and 4200 R, the reduction in bone deposition and osteoradiolysis in the mandibles of animals given large doses of ascorbic acid was less pronounced than in animals receiving small doses despite the similarity in preirradiation bony structure in the two groups. Animals given small doses of ascorbic acid and 2600 or 4200 R had odontoblasts that were still active in the production of normal dentin. Those subjected to low doses of ascorbic acid and 800 to 1200 R showed increased disturbances in function of the scorbutically damaged odontoblasts. At doses of 2600 and 4200 R, there was total disintegration of odontoblasts and pulp cells, accompanied by circulatory disruptions in the pulp. Animals injected with large doses of ascorbic acid had osteoblasts that were more radiosensitive than osteocytes or mature odontoblasts. At the low doses of ascorbic acid, the scorbutically changed odontoblasts were more radiosensitive than osteoblasts or osteocytes.

Vitamin C Studies in Man

Hodges et al.[127] produced scurvy in four prisoner volunteers given a liquid formula diet by self-administered stomach tube that was totally deficient in ascorbic acid. The diet was adequate in all other nutrients and was consumed while the volunteers were housed in a metabolic ward. The vitamin C-free formula diet was ingested for 114 days. Repletion of ascorbic acid was begun on day 100 at a daily level of 4 mg for the first subject, 8 mg for the second, 16 mg for the third, and 32 mg for the fourth. On the 14th day of repletion (day 113), the volunteers were switched from the formula diet to a soft diet that provided 2.5 mg ascorbic acid per day. All subjects had their natural teeth with varying degrees of periodontal inflammation.

In the first subject, gingival hemorrhages appeared on day 43 in association with periodontal inflammation. The second, who had swollen bleeding gums to start with, developed an increase in swelling on day 32. By day 36, gums were intensely congested and engorged; by day 91, there was pronounced swelling of the interdental papillae. Within 1 month of repletion with 10.5 mg ascorbic acid daily, the gingiva returned to the original condition. The third subject developed redness of the gingival margins on day 105 and swelling around the molars a few days later. His gingiva appeared normal on day 180 while on a daily intake of 6.5 mg ascorbic acid. The fourth subject had slight bleeding of the gums on day 83 without further intensification. In all instances, the gingival lesions began along the margins and later involved the interdental papillae. Small sublingual petechial hemorrhages were noted in three subjects beginning on days 36, 90, and 189, respectively. These cleared in 1 to 4 days following ascorbic acid therapy.

Vitamin D Studies in Animals

Weinmann and Schour[128] examined the teeth of 33 white rats fed a rachitogenic diet high in calcium, low in phosphorus, and deficient in vitamin D for 1 to 56 days after weaning. Serving as controls were 25 litter mates. Histologically, enamel formation and calcification were normal. The enamel organ showed cystic degeneration in the incisal half of the incisor. Dentin formation was retarded in rate and calcification. The immediate response was the deposition of a calciotraumatic line. The newly formed dentin had an interglobular texture. Enamel-covered dentin was more severely affected by the deficiency than cementum-covered dentin. Cementum formation was normal in rate, but calcification was defective. The rachitic animals had a much higher frequency of shallow areas of molar root resorption than the control animals.

The same rats were also used by Weinmann and Schour[129] to assess the effects of a rachitogenic diet on alveolar bone. Histologically, the formation of new bone in the rachitic rats proceeded at a normal rate, but remained uncalcified and persisted as osteoid tissue. Osteoid failed to undergo resorption, resulting in excessive accumulation. Sites that normally show apposition were undisturbed; those which normally show resorption were inactive. The imbalance resulted in a distortion of the growth pattern of the alveolar bone. There was a reduction in periodontal ligament space in some areas. Compression of the periodontal tissues led to hyaline degeneration of the connective tissue and, in some instances, to complete obliteration of the periodontal ligament.

The action of parathyroid hormone on the alveolar bone and teeth of rachitic rats was reported by Weinmann and Schour.[130] In this study, 38 white rats placed on a rachitogenic diet after weaning and 23 animals on a normal diet were given single or multiple injections of parathyroid hormone at age 48 to 52 days. Doses varied from 1 to 24 in number and from 50 to 780 Hanson units in quantity. The animals were killed from 1 to 132 hr after the last injection. Sixteen normal and 17 rachitic rats not given parathyroid were used as controls. Immediately after sacrifice, the heads were removed and fixed in 10% formalin. The jaws were dissected, upper incisor and molar areas prepared for histologic section, and stained with hematoxylin and eosin.

Injections of 50 Hanson units of parathyroid hormone stimulated osteoblastic activity and formation of new alveolar bone. Injections of 100 or more Hanson units induced progressive osteoclasis and replacement of bone by fibrous tissue. Bone reaction to parathyroid hormone was essentially the same in the normal and rachitic rats except that the osteoid tissue in rachitic animals was more resistant to resorption. Bone in the molar areas was more sensitive to the hormone than bone in the incisal area. Injection of parathyroid hormone did not cause any resorption of enamel, dentin, or cementum in any of the animals, regardless of nutritional status.

Since starvation has the same healing effect on rachitic epiphyseal cartilage as suitable amounts of cod liver oil, Weinmann and Schour[131] extended these studies to the dentin of rachitic rats. Incisal dentin from four albino rats on a rachitogenic diet for 4 weeks and given 3 m*l* of irradiated ergosterol 1 to 4 days before sacrifice and from four litter mates on the same diet for the same time, but starved for 1 to 4 days before termination, was studied histologically. Both fasting and vitamin D caused identical reparative changes in the dentin of the rachitic rats. Calcification of the wide rachitic predentin began in the oldest layer and was best seen in the basal predentin. In each instance, the layer of intermediate dentin that disappeared during rickets redeveloped in the basal third of the enamel-covered dentin.

Weinmann[132] described the rachitic changes in the mandibular condyle of the rat. The mandible grows at the condyle in a manner similar to the growth of tubular bone at the epiphyseal plate and articular cartilage. In this study, 14 white rats were fed a rachitogenic diet for 7 weeks after weaning. The animals were then sacrificed and the heads removed and fixed in Zenker-formalin solution. After decalcification, the temperomandibular joint was cut into blocks, processed, sectioned, and stained with hematoxylin and eosin. Slides from litter mate control animals were prepared in the same manner.

The effects of the rachitogenic diet on the mandibular condyle were analogous to those in the articular and epiphyseal cartilages of tubular bone. Growth cartilage in the condyle increased in thickness and the condyle per se was club-shaped. Both changes were attributable to a lack of calcification of cartilage and osteoid tissue and resistance to resorption, preventing replacement by bone. Bone formed during the rachitogenic regimen remained uncalcified and persisted as osteoid tissue. Normal reconstruction and modeling resorption were arrested, leading to increased density of the spongiosa and a clublike thickening of the condyle.

Durkin et al.[133] compared the changes in the articular, mandibular condyle, and growth plate cartilages during the onset and healing of rickets in rats. In this study, 81 Holtzman rats, 23 days of age, were confined to a rachitogenic diet; 17 of these received 30 IU vitamin D per animal per day by mouth and constituted the controls. Animals from each group were sacrificed every day from day 0 to day 18 and also on days 21 and 28. Specimens were processed for histologic examination of the left mandibular condyle, distal head of the femur, and proximal head of the tibia.

Rachitic changes in the articular and condylar cartilages were manifested by an increase in width confined entirely to the zone of hypertrophic cartilage. Rachitic changes in the tibial growth plate cartilage were more severe than in the other two. Loss of normal relationship between columns of chondrocytes and metaphyseal vessels resulting from removal of large irregular areas of cartilage was present in the rachitic tibial growth plate. This was not seen in the tibial articular or mandibular condylar cartilages.

After vitamin D administration, the hypertrophic type of cartilage in the tibial articular cartilage was rapidly reduced in width and transformed into the nonhypertrophic type. Rachitic cartilage in the condyle was rapidly narrowed and restored to normal by the 10th day. The development of calcification was similar to that during the early postembryonic period. Calcification of the growth plate during healing resembled for a short time the pattern in the tibial anlage and condylar and articular cartilages. The rachitic disorder has a retrograde effect on the cellular characteristics of the tibial growth plate and makes the inherent embryonic nature of the articular and condylar cartilages more prominent.

Dreizen et al.[134] studied the interrelationships between vitamin D deficiency osteomalacia and hyperparathyroidism in nine adult marmosets evenly distributed among *Saguinus oedipus, S. fuscicollis,* and *Callithrix jacchus.* Because the vitamin D requirements of the marmoset are both uniquely high and structurally specific, this primate is an excellent experimental model for the study of vitamin D deficiency. In the present investigation, one member from each species received a Vitamin D_3-deficient fruit and vegetable diet and the others the same diet fortified with 500 IU vitamin D_3 per animal per day. At the time of death, each of the vitamin D_3 deficient animals had clinical and radiologic manifestations of osteomalacia. This was evidenced by a partial paralysis of the lower extremities, atrophy of the muscles of the loin and limbs, marked bowing of the spinal column with compression fractures of the thoracic and lumbar vertebrae, pathologic fractures and distortions of the long bones, and extensive skeletal demineralization.

Hyperparathyroidism was induced in three of the marmosets fed the vitamin D_3-containing diet by twice daily injections of 100 units parathyroid USP suspended in 0.8% gelatin solution. One animal died on the third day after receiving a total of 400 units, the other two on the 5th day after a total of 800 units. The control group was comprised of the remaining three marmosets on the vitamin D_3 supplemented ration. Each eventually succumbed to acute pneumonitis, but was free from detectable bone disease at the time of expiration. At necropsy, the tissues were fixed in 10% formalin, the skulls sawed midsagittally and grenz-rayed. Fixed jaws were cut into blocks, decalcified, embedded in paraffin, sectioned, and stained with hematoxylin and erythrosin B, PAS and hematoxylin, Wilder's silver stain, and van Gieson's picrofuchsin.

Radiographs of the skulls of marmosets with osteomalacia and with experimentally induced hyperparathyroidism revealed strikingly similar derangements in bone structure. In each, there was a generalized partial to complete disappearance of the lamina dura and a pronounced diminution in density of the compact and cancellous bone elements. The overall pattern of extensive demineralization was represented by greatly increased radiolucency of the trabecular interstices, loss of trabeculae, breaks in trabecular continuity, increased prominence of the remaining trabeculae, and reduction in width of the labial, buccal, and lingual cortical plates of the alveolar process.

Histologically, the osteomalacic animals showed bone changes ranging from a marked thinning and partial loss of the cribriform plates, cortical plates, and cancellous lamellae to an almost complete absence of these structures. Some of the remaining lamellae contained osteoid seams surrounded by a single layer of osteoblasts; around others there was extensive osteoclastic activity with formation of Howship's lacunae. Along the entire depth of the alveolar process the bone marrow was replaced by proliferating fibrous tissue containing numerous fibroblasts and comparatively little collagen. Bone loss in the parathyroid-injected animals was severe, with almost all of the compact and cancellous bone supporting the teeth destroyed by excessive osteoclastic resorption. Interdental spaces were studded with multinucleated giant cell osteoclasts, and the residual bone was peppered with Howship's lacunae. Progression of lacunar absorption was accompanied by a proliferation of fibrous tissue and bone marrow fibrosis. Osteoblasts and narrow osteoid seams surrounded remnants of unresorbed trabeculae. Basic similarities between the osteomalacic and parathyroid-injected animals suggested that secondary hyperparathyroidism activated by osteomalacia may have contributed to the bone changes in the vitamin D deficient marmosets (Figure 8).

Fahmy et al.[135] elicited the effects of hypervitaminosis D on the hamster periodontium. Excess vitamin D has been reported to produce a variety of modifications in the human oral tissues. Rampant dental caries, malocclusion, rarefaction of molar roots, and thinning of enamel, dentin, and alveolar bone have been found in children with hypervitaminosis D. In rats, this condition produces decreased incisor tooth growth, pulp stones, hypercementosis, ankylosis due to thickened cementoid, alterations in pulpal dentin, hemorrhage into the periodontal membrane, and atrophy of the enamel organ. In dogs, the manifestations include decreased jaw size, sclerosis of the jaws, and calcified deposits in the periodontal membrane and free gingiva.

In the present study, 30 male and 30 female hamsters approximately 4 months old were arranged into experimental, pair-fed control and *ad libitum* fed control groups balanced in sex and litter origin. Hamsters in the experimental group received i.p. injections of 0.01 mℓ sesame oil containing 5000 units vitamin D_2 twice a week. Each received a total of 16 injections containing 80,000 units vitamin D_2. All were sacrificed on the 56th study day. The tissues were fixed in 10% formalin except for the salivary glands, which were fixed in Bouin's. Half the mandible was decalcified, blocked, embedded in paraffin, and selected sections stained with hematoxylin and eosin and with von Kossa.

All experimental animals developed hypercalcemia and had diminished femur ash content. They also had varying degrees of calcification in the kidneys involving the tubules, blood vessel walls, interstitial tissues, glomeruli, and vascular spaces in the cortex. Histologic changes in the femur were restricted to the cortical and marrow

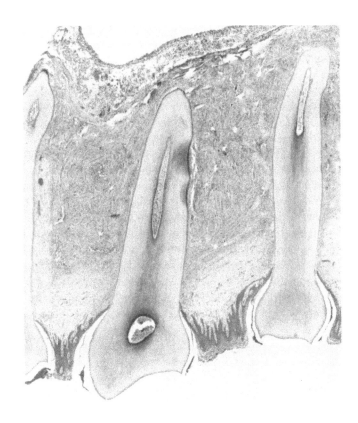

FIGURE 8. Periodontal bone destruction and alveolar osteitis fibrosa in marmoset with vitamin D deficiency osteomalacia. (Hematoxylin and erythrosin B stain; magnification × 20.) (From Dreizen, S., Levy, B. M., Bernick, S., Hampton, J. K., Jr., and Kraintz, L., *Isr. J. Med. Sci.*, 3, 731, 1967. With permission.)

portions of the diaphysis. Principal abnormalities included numerous excavations on the endosteal side of the diaphysis and irregular outgrowths from the cortex into the marrow cavity. The excavations were filled with marrow and lined in places with osteoid. Overall thickness of the shaft was decreased. In the molar regions, the periodontium showed central resorption of interdental and interradicular alveolar bone. Irregular, marrow-filled spaces were present that occasionally contained considerable osteoid. Identical changes were found in the central part of the mandible. Bone adjacent to the cribriform plates had basophilic lines of Pagetoid appearance suggestive of alternate bone formation and resorption. Osteoid covered the interdental crests and lined the socket wall creating an impression of ankylosis, but there was no fusion between cementum and bone. The principal fibers of the periodontal membranes stained deeply basophilic in the region of the cementum. Some basophilic fibers appeared fused. Sharpey's fibers also stained dark, suggestive of calcification.

Vitamin E Studies in Animals

Granados et al.[136] discovered that rearing Florida cotton rats, albino rats, and hamsters from weaning until 5 to 6 months of age on a vitamin E-deficient diet containing 10% lard and 2% cod liver oil produced complete depigmentation of the maxillary incisors in the albino rats and hamsters, but not in the cotton rats. Ability of the vitamin E-deficient diets to inhibit enamel pigmentation in rodents was related to the fat component of the diet. Histologically, albino rat jaws from animals in which cod liver oil replaced lard in the vitamin E-deficient diet, showed progressive atrophy of the enamel organs and increasing deposition of an acid-fast pigment in the bone marrow fat, and in macrophages surrounding the enamel organs. Identical histologic changes were seen in albino rats when cod liver oil was replaced by highly unsaturated fatty acids. Feeding of medium and low unsaturated fats produced no such changes nor did a vitamin E-deficient diet devoid of fat. Vitamin E therapy progressively restored enamel organ function and enamel pigment deposition, but did not appreciably influence pigment formation in the adjacent periodontal tissues.

Pindborg[137] made an exhaustive study of the effects of avitaminosis E on the enamel organs of rat incisors. Five groups of 1-month-old Wistar rats were used in the experiment. Group I had 120 rats that received a vitamin E-deficient purified diet for 23 to 425 days. Group II encompassed 28 rats fed the same diet as Group I supplemented with 3.5 mg tocopherol per rat per day for 34 to 221 days. Group III contained 10 rats fed a fat-free vitamin E deficient diet for 66 to 119. Group IV consisted of 15 rats given the same diet as Group III for 66 to 119 days in which 20% sucrose was replaced by 20% cod liver oil. Group V included 10 rats given a diet for 3 to 10 months that was made deficient in vitamin E by treatment with ferric chloride. At sacrifice, the teeth were examined macroscopically for evidence of depigmentation, fixed in formalin, decalcified, embedded in celloidin, sectioned, and stained with hematoxylin and eosin.

Rats fed three of the four vitamin E-deficient diets showed lack of pigmentation in the upper incisors of varying degrees, whereas the lower incisors had only occasional slight changes in color. Those given the fat-free vitamin E-deficient diet had no changes in the upper incisors and only rare changes in the lower incisors. All animals given tocopherol supplements were protected from vitamin E deficiency-related manifestations. Each rat on the vitamin E-deficient diets except the fat-free variety had abnormalities in the enamel organs of the upper incisors consisting of damage to the capillary walls in the papillary layer, edema in the papillary layer, folding of the ameloblasts into the adjacent edematous connective tissue with cyst formation, and change in the timing of atrophy of the papillary layer and ameloblasts.

Nelson and Chaudry[138] studied the effects of avitaminosis E on the oral, paraoral, and hematopoeitic tissues of the rat. In this study, 59 30-day-old male albino Wistar rats were divided into three groups. Group I (31 animals) was fed a totally tocopherol-deficient test diet *ad libitum* for 7 to 14 months. Group II (14 animals) was given the same diet *ad libitum* for 14 months plus 10.215 g tocopherol per 100 lb of diet. Group III (14 animals) received the same diet as Group II and was pair-fed with Group I. After 14 study months, the rats were sacrificed with chloroform, heads split midsagittally, and specimens of the salivary glands, lips, tongue, buccal muscles, and decalcified teeth and jaws were processed for histologic examination.

The vitamin E-deficient rats developed progressive muscular dystrophy and weight loss. Oral musculature exhibited loss of the characteristic striations, fatty changes, fibrosis, and swelling and hyalinization of muscle fibers. Degeneration of muscle nuclei was a constant finding. Salivary glands of the avitaminotic E rats underwent replacement of parenchyma by connective tissue that was most pronounced in the parotid gland. Fatty changes were exhibited mainly by the submandibular gland. There was a marked retardation in the eruption rate of the incisor teeth. Depigmentation of these teeth was evident at the end of four study months. The teeth were brittle and hypoplastic. Pigmentation commonly associated with vitamin E deficiency was present in the soft tissues of the oral cavity and in the salivary glands. There were no significant histologic changes either in the periodontium or dental pulp.

INANITION STUDIES IN ANIMALS

Tonge and McCance[139] determined the influence of severe undernutrition on the mouth structures of growing and adult pigs. The animals had been undernourished for 52 weeks. Controls were pigs either of the same weight but much younger, or of the same age but much larger. Twelve undernourished pigs were killed at 1 year of age; six normals, at 4 weeks of age (weight or size controls); four normals, at 52 weeks of age (age controls). Jaws and soft tissues removed at necropsy were fixed in formalin, examined grossly, and processed for histologic study.

Severe undernutrition altered the shape and anatomic relationships of the jaws; retarded normal growth of the gingiva and palatal ridges; delayed development of the teeth less than that of the jaws, resulting in overcrowding, displacement and malocclusion; prolonged retention of the milk teeth; altered the structure of both enamel and dentin, and probably reduced the size of the teeth formed during the period of malnutrition; altered the X-ray appearance and structure of the maxilla and mandible producing multiple "lines of arrested growth".

Tamarin et al.[140] examined the relative reactivity of the major sublingual gland, submaxillary gland, and exocrine pancreas of the rat to total starvation. The sublingual gland in the rat is mainly mucus producing; the pancreas, completely enzyme producing (serous); the submaxillary gland, mixed mucous and serous. Young mature male Sprague-Dawley rats with a mean body weight of 333 g were used. All animals underwent a preliminary starvation period of 9 hr before feeding for 2 hr. Food but not water was then removed for periods of up to 9 days. The first six animals were killed after 3 to 216 hr of starvation. Histometric and histomorphic evaluations were made on all animals. Slides were stained with hematoxylin and eosin, Azure blue and methylene blue, PAS, and Alcian blue. Stains for trypsinlike esterases, protease, and tryptophan were also used. Protease, amylase, sialic acid, RNA, and DNA were measured from frozen tissue that had been lyophilized.

The pancreas lost weight at a faster rate than either salivary gland or total body during the various periods of starvation. Each organ had an increase in number of cells per unit area. Rate of acinar cell increase was significantly greater in the pancreas than in the salivary glands. Parenchymatous units of all three organs were reduced in size during starvation, with the greatest structural alterations in the acini. After 9 days of starvation, the salivary glands demonstrated changes in acinar components that were more prominent in the submaxillary than in the sublingual gland.

All glands had an increase in DNA per milligram wet tissue and a decrease in RNA per unit of DNA. The synthetic mechanisms of all three organs were steadily reduced during starvation. The reductions were apparent in amylase, protein, and mucin production. During starvation, the body follows a unidirectional downhill course. The effects are not the same in every organ and tissue. Distinct priorities for survival exist. Cell vitality as reflected by decreasing organ weights, increasing numbers of cells per unit area, increasing DNA per weight, decreasing total DNA per whole organ indicate that the pancreas is more labile and more responsive to starvation than the sublingual and submaxillary salivary glands.

FAT STUDIES IN ANIMALS

In the course of testing whether marmosets could be made vulnerable to atherosclerosis by dietary means, Dreizen et al.[141,142] produced a unique malabsorption syndrome manifested by jejunal lipodystrophy, steatorrhea, and osteomalacia in four adult cotton top marmosets *(S. oedipus).* The animals were restricted to a purified high choles-

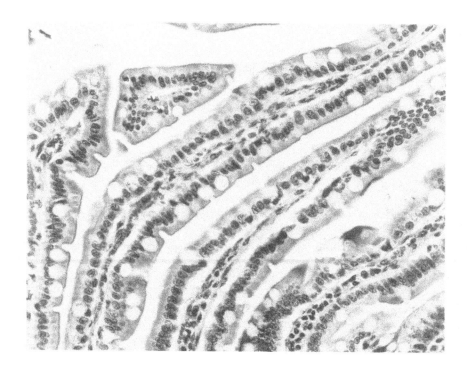

FIGURE 9A.   Histologic appearance of jejunal villi in normal marmoset. (Hematoxylin and erythrosin B stain; magnification × 400.) (From Dreizen, S., Levy, B. M., and Bernick, S., *Proc. Soc. Exp. Biol. Med.,* 138, 7, 1971. With permission.)

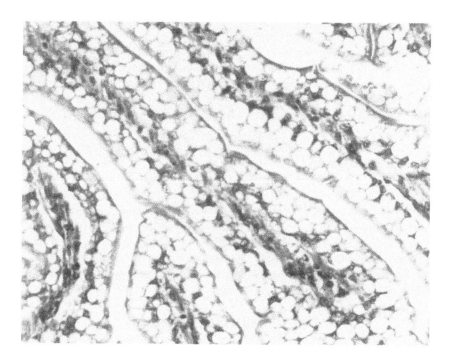

FIGURE 9B.   Lipodystrophy in jejunal villi of marmoset consuming high cholesterol, high coconut oil diet for 42 weeks. (Hematoxylin and erythrosin B stain; magnification × 400.) (From Dreizen, S., Levy, B. M., and Bernick, S., *Proc. Soc. Exp. Biol. Med.,* 138, 7, 1971. With permission.)

teral (5%), high coconut oil (23%) diet that included adequate amounts of all nutritional essentials for this species. All the marmosets developed striking structural changes in the small intestine that were confined to the jejunal villi and spared the duodenal and ileal counterparts. The cytoplasmic component of the absorptive cells lining the villi was displaced by large, clear, fat-containing vacuoles that distended the cell, compressed the nucleus, and made each cell indistinguishable from adjoining goblet cells. The malabsorption induced osteomalacia involved the entire skeleton and was productive of pathologic fractures in the long bones of the extremities (Figures 9A and 9B).

Radiographically, there was a thinning of the cortical bone with marked lamellation, cortical microfractures, coarse and uneven trabecular networks, and widening of medullary canals. In each skull there was a partial to complete loss of lamina dura and greatly increased patchy radiolucencies in the mandible, maxilla, and calvarium. Histologically, the alveolar bone changes were denoted by a narrowing and disruption of the cortical and cribriform plates and by the presence of broad bands of osteoid around the remaining compact and spongy bone elements.

Dreizen et al.[143] produced atherosclerosis in the oral arterial vasculature of rabbits by confining 20 male New Zealand white rabbits weighing 1.25 kg to a diet composed of 96.75% Purina® rabbit chow, 3.00% Wesson pure vegetable oil, and 0.25% cholesterol. The rabbits were distributed into four groups of five each. Group I received deionized water for drinking, and Groups II, III, and IV received deionized water plus 1 ppm F, 10 ppm F, and 0.01% d-catechin, respectively. A fifth group maintained on regular chow and drinking water served as controls. Food and water were provided ad libitum. One rabbit from each group was killed by cardiac puncture exsanguination at the end of the second, third, fourth, fifth, and sixth study month. At necropsy, the heads were removed, fixed in formalin, cut into blocks, decalcified, paraffin-embedded, sectioned, and stained with hematoxylin and erythrosin B or Masson's trichrome.

All animals on the cholesterol-supplemented diet, irrespective of the composition of the drinking water, developed extensive atherosclerosis in the arteries of the tongue, lips, buccal mucosa, palatal mucosa, gingiva, periodontal ligament, periodontal bone, major and minor salivary glands, and skeletal muscles of the face. Arteries of large, medium, and small caliber contained intimal collections of foam cells. Lumens of many of the small arteries and arterioles were almost completely obliterated by the atheromatous incursions, but none of the lesions was complicated by thrombosis. Histologic integrity of the oral structures was not detectably disturbed by the vascular reaction to the atherogenic diet (Figures 10A and 10B).

Although the foam cell accumulations resembled somewhat the fatty streak stage of human atherosclerosis, they did not progress to advanced lesions so long as the rabbits were kept on the atherogenic ration. Accordingly, Dreizen et al.[144] extended the studies to determine whether repetitive alteration of the atherogenic diet with a nonatherogenic diet would promote formation of more advanced lesions. Six New Zealand white rabbits 6 to 7 weeks of age were given the previously described atherogenic diet every odd month and a nonatherogenic diet comprised completely of Purina® rabbit chow every even month for 2 years. Rabbits of comparable strain and age given the nonatherogenic diet throughout the study were used as controls. All were killed by cardiac puncture exsanguination after 24 months. The tissues were processed as in the previous study and examined microscopically.

Interval cholesterol feeding protocol was productive of advanced intimal atherosclerosis in the form of fibro-fatty plaques and of medial arterioslcerosis in the aorta and many visceral and peripheral arteries, notably, the lingual, coronary, pulmonary, gastric, and renal arteries. None of the lesions was completely occlusive or infarct-producing. Intraorally, intimal, and medial arteriopathies uncomplicated by ulceration, hem-

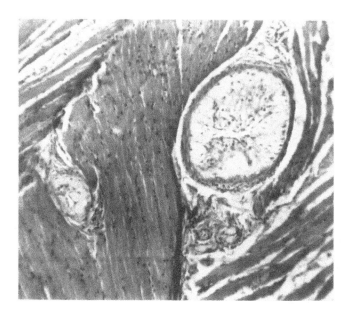

FIGURE 10A.   Foam-cell atheromas in central branches of lingual artery of rabbit fed atherogenic diet for 4 months. (Masson's trichrome stain; magnification × 100.) (From Dreizen, S., Vogel, J. J., and Levy, B. M., *Arch. Oral Biol.*, 16, 43, 1971. With permission.)

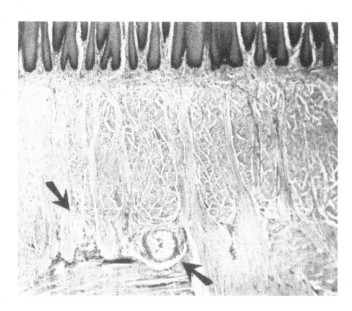

FIGURE 10B.   Foam-cell atheromas in peripheral branches of lingual artery in rabbit given atherogenic diet for 4 months. (Masson's trichrome stain; magnification × 43.) (From Dreizen, S., Vogel, J. J., and Levy, B. M., *Arch. Oral Biol.*, 16, 43, 1971. With permission.)

orrhage, or thrombosis were manifested by the labial, gingival, palatal, periodontal, and alveolar arteries, in addition to the lingual arteries. The histologic structure of tissues supplied by these arteries was not affected by the experimentally produced arteriopathies (Figures 11A and 11B).

Wirthlin et al.[145] fed 20 New Zealand white rabbits, 8 weeks of age, a diet of dry pelleted rabbit chow and water *ad libitum*. Ten animals also received a daily supplement of 1 g cholesterol suspended in 5 m*l* corn oil 6 days a week for 6 months. All were subjected to a standard injury created by insertion of ligature wire through the interdental space below the contact area between the lower right first and second premolars. Animals given cholesterol developed hypercholesteremia and marked fatty lesions in the walls of the aortic arch, fatty metamorphosis of the liver and scattered foci of lipophages in the oral mucosa, skin, lungs, spleen, and liver. The mandibular arteries in the cholesterol-fed group had slightly narrower average diameters and slightly thicker arterial walls. Atherosclerotic lesions typical of cholesterol feeding were found in the lingual arteries. The interdental ligatures caused recession of the gingival papillae and a significant loss in the height of the interbony septae, as measured on radiographs. Bone height loss was not aggravated by the cholesterol feeding.

Different species of New World monkeys differ markedly in their vulnerability to both spontaneous and experimentally induced atherosclerosis. Systematic surveys of wild caught specimens have demonstrated that naturally occurring atheroslcerosis is rare in marmosets. The pronounced resistance of the marmoset to atherosclerosis in the wild prompted Dreizen et al.[146] to attempt to produce the disease in captive animals restricted to a high cholesterol-high lard diet. Four adult, healthy, colony-conditioned cotton top marmosets *(S. oedipus)* were confined to a nutritionally adequate chemically defined diet containing 5% cholesterol, 23% lard, and 2% corn oil. Serum cholesterol levels were determined from samples of femoral vein blood drawn just before the start of the experimental regimen and at monthly intervals thereafter. They were killed by cardiac puncture exsanguination when the cholesterol concentrations approximated 4, 6, 8, and 16 times the initial values. These levels were reached at 47, 52, 68, and 73 weeks, respectively. All structures removed at necropsy were formalin fixed, processed for paraffin-embedding, sectioned, and alternate slides stained with hematoxylin and erythrosin B, Masson's trichrome, and Weigert's iron hematoxylin. Selected specimens of formalin-fixed hearts, aortas, and tongues were embedded in gelatin, sectioned in a freezing microtome, and stained with Oil Red O.

Grossly, demonstrable sudanophilia indicative of intimal fat deposition was found only in the aortas of marmosets followed for more than 1 year and not in those killed at 48 and 52 study weeks. Microscopically, each marmoset had prominent atherosclerotic changes in the lingual arteries and their branches. All of the arteriolar — and the vast majority of the arterial — lesions were of the histiocytic foam cell variety. In some of the medium-sized branches of the lingual arteries, the lesions consisted of intimal foam cells and lipid laden multipotential mesenchymal cells. An unusual and provocative aspect of the study was that atherosclerosis onset in the lingual arteries preceded the disease in the aorta and contiguous major arteries (Figures 12A and 12B).

Doyle et al.[147] kept four *Macaca irus* monkeys on controlled diets for slightly more than 1 year. Two received a high fat, high cholesterol ration; two, a standard diet. Since the monkeys would not eat the test diet every day, an alternate regimen consisting of two oranges and two apples cut and mixed with Purina® monkey chow was given on a random schedule. Both monkeys on the atherogenic diet developed the disease in the oral arterial vasculature. The mildest lesion was represented by a slight intimal thickening and splitting and reduplication of the internal elastic lamina. Isolated large eccentric fibrous plaques occluding from 33 to 90% of the lumen were found in the inferior alveolar artery and in small arteries supplying the nasal mucosa and mandi-

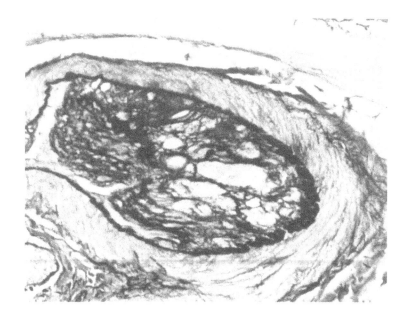

FIGURE 11A. Extensive fibro-fatty atheroma in lingual artery of rabbit on intermittent cholesterol feeding for 2 years. (Aldehyde fuchsin stain; magnification ×
400.) (From Dreizen, S., Stern, M. H., and Levy, B. M., *J. Dent. Res.,* 57, 412,
1978. With permission.)

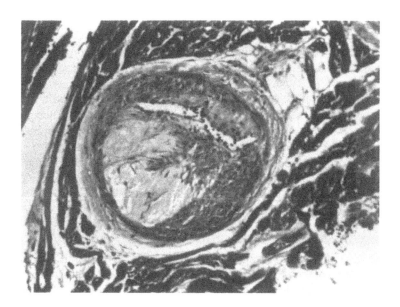

FIGURE 11B. Extensive fibro-fatty atheroma and medial sclerosis in coronary
artery of rabbit on intermittent cholesterol feeding for 2 years. (Masson's trichrome
stain; magnification × 100.) (From Dreizen, S., Stern, M. H., and Levy, B. M., *J.
Dent. Res.,* 57, 412, 1978. With permission.)

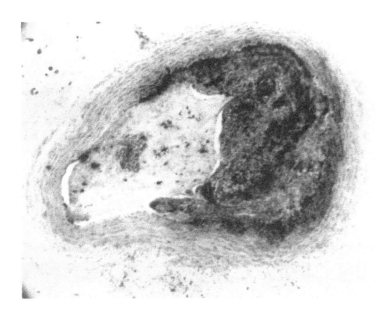

FIGURE 12A. Atheromatous lesion in frozen section of abdominal aorta of marmoset fed atherogenic diet for 73 weeks showing heavy intimal accumulation of fat-filled foam cells. (Oil red O stain; magnification × 20.) (From Dreizen, S., Levy, B. M., and Bernick, S., *Proc. Soc. Exp. Biol. Med.*, 143, 1218, 1973. With permission.)

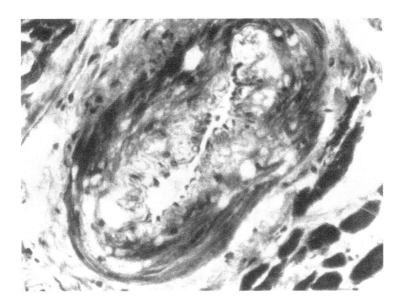

FIGURE 12B. Atherosclerotic changes in intima and media of major branch of lingual artery in marmoset relegated to the atherogenic diet for 68 weeks (Masson's trichrome stain; magnification × 50.) (From Dreizen, S., Levy, B. M., and Bernick, S., *Proc. Soc. Exp. Biol. Med.*, 143, 1218, 1973. With permission.)

bular molars. Whether any of the partial occlusions impaired the nutrition of the periodontium in the atherosclerotic monkeys was not apparent histologically.

Alveolar bone loss in rats given an atherogenic diet was reported by Plenk et al.[148] Male Wistar rats aged 4 and 6 months were fed a combination of 52.7% stock diet, 40% lard, 5% cholesterol, 2% sodium cholate, and 0.3% 2-thiouracil. Six rats were fed this diet for 6 weeks; six, for 8 weeks. Controls were comprised of animals on the stock diet and on stock diet plus thiouracil for 6 and 8 weeks. All rats were killed at the end of the corresponding feeding periods. The heads were fixed in formalin, the upper jaws sawed into blocks, decalcified, paraffin-embedded, sectioned, and alternate slides stained with hematoxylin and eosin, PAS-Alcian blue, and Mallory's trichrome. Remodeling of the alveolar bone was studied histologically in the alveolar walls of the incisor and molar teeth.

Diminished formation of new bone and increased osteoclastic activity were found in the alveolar bone of rats on the atherosclerotic diet. The changes were more pronounced in the incisal than in the molar area. Animals on the atherogenic diet also manifested gingival inflammation and periodontal pocket formation. There were no vascular changes relatable to the diet in these animals.

LATHYRISM STUDIES IN ANIMALS

Krikos et al.[149] described the oral changes in experimental lathyrism in rats. Diets containing sweet pea seeds *(Lathyrus odoratus)* produce a disease in this rodent that primarily affects the connective tissues. The culpable chemical is β-aminoproprionitrile, a two-carbon compound with an amino group on one end and a nitrile group on the other.

In this study, 70 male and female Wistar rats, 21 days of age, were divided into two groups. Group I was fed a synthetic diet *ad libitum*; Group II, the same diet containing 0.2% β-aminoproprionitrile. Animals from both groups were killed at intervals between 3 and 28 days on the diets. Oral cavities were examined grossly at the time of sacrifice and one half of the mandible was processed for histologic examination.

Mandibles from the lathyritic animals showed exostoses in the region of the tendonous attachments typified by a thick, highly cellular periosteum that covered bone composed of thin trabeculae and broad marrow spaces. The lathyritic rats had increased mobility of the teeth. Periodontal ligament fibroblasts exhibited increased cytoplasmic basophilia and palisading. Collagen fibers were fine, disoriented, and embedded in amorphous eosinophilic material. Condylar cartilage was enlarged with a matrix that contained irregularly outlined foci filled with granular material. Cartilage cells were arranged in isogenous groups and failed to undergo enlargement and vacuolization.

Abramovich[59] investigated the changes in the facial structures of rats caused by dietary supplements of *L. odoratus*. In this study, 16 female white rats were mated and the date of conception established, after which they were separated into two groups of 8 each. Group I was fed a standard diet; Group II, a diet containing 50% *L. odoratus*. The fetuses were removed by Caesarean section on the 18th day of gestation, fixed in formalin, paraffin-embedded, and sectioned frontally.

Micrometric measurements were made on stained sections through the anterior region at a plane that passed through Jacobson's organ and in the posterior region at a plane that passed through the first molar germs. In the anterior region, measurements were made of the height of the head, height of the nasal septum, and height of the cartilage in the nasal septum; in the posterior region, measurements included width of the tongue at the lingual groove and height of the tongue from the dorsum to the geniohyoid muscle. The sweet pea seed diet produced cleft palates in almost all of the fetuses. *L. odoratus* inhibited development of the palatal shelves and caused the appearance of cartilagenous masses at the level of the first mandibular molar. Although the heights of the face and total nasal septum were not significantly different from the controls, the lathyritic fetuses had a significantly reduced height of nasal septum cartilage and increased thickness of the mandible.

As evidenced by the foregoing, formation, maintenance, replacement, and repair of the mouth tissues are nutrition dependent. Oral reactions to nutritional disturbances are characterized by an interference with growth and maturation of the epithelial, fibrillar, osseous, and/or dental elements expressed clinically as a degeneration of the formed structures. These changes also contribute to a loss of the tissue barriers to infections, with attendant inflammatory and disruptive sequelae. Among the nutrients which have been shown to affect tissue integrity in the oral cavity under experimental conditions are protein, calcium, phosphorus, magnesium, zinc, copper, iron, vitamin A, riboflavin, niacin, pyridoxine, folic acid, biotin, vitamin B_{12}, vitamin C, vitamin D, and vitamin E.

REFERENCES

1. Stein, G. and Spencer, H., Changes of the human tongue in protein and vitamin depletion and repletion, *Ann. N.Y. Acad. Sci.*, 85, 368, 1960.
2. Squires, B. T., Differential staining of buccal epithelial smears as an indicator of poor nutritional status due to protein-calorie deficiency, *J. Pediatr.*, 66, 891, 1965.
3. Squires, B. T., Buccal mucosa in protein-calorie deficiency in the pig, *Br. J. Nutr.*, 17, 303, 1963.
4. Stahl, S. S., Sandler, H. C., and Cahn, L., The effects of protein deprivation upon the oral tissues of the rat and particularly upon periodontal structures under irritation, *Oral Surg. Oral Med. Oral Pathol.*, 8, 760, 1955.
5. Person, P., Wannamacher, R., and Fine, A., The response of adult rat oral tissues to protein depletion: histologic observation and nitrogen analysis, *J. Dent. Res.*, 37, 292, 1958.
6. Ziskin, D. E., Karshan, M., and Stein, G., Epithelial changes of the tongue in rats fed a protein-deficient diet, *J. Dent. Res.*, 28, 732, 1948.
7. Johnson, D. A., Sreebny, L. M., and Enwonwu, C. O., Effect of protein-energy malnutrition and of a powdered diet on the parotid gland and pancreas of young rats, *J. Nutr.*, 107, 1235, 1977.
8. Pack, A. R. C., A histometric evaluation of alveolar trabecular bone density in the jaws of protein-deficient rats, *Arch. Oral Biol.*, 23, 871, 1978.
9. Chawla, T. N. and Glickman, I., Protein deprivation of the periodontal structures of the albino rat, *Oral Surg. Oral Med. Oral Pathol.*, 4, 578, 1951.
10. Goldman, H. M., The effects of dietary protein deprivation and of age on the periodontal tissues of the rat and spider monkey, *J. Periodontol.*, 25, 87, 1954.
11. Shaw, J. H., Influence of marginal and complete protein deficiency for varying periods during reproduction on growth, third molar eruption and dental caries in rats, *J. Dent. Res.*, 48, 310, 1969.
12. Shaw, J. H., Marginal protein deficiency during the reproductive cycle in rats: influence on body weight and development of skulls and teeth of offsprings, *J. Dent. Res.*, 49, 350, 1970.
13. Stahl, S. S., Healing of gingival wounds in female rats fed a low protein diet, *J. Dent. Res.*, 42, 1511, 1963.
14. Stahl, S. S., Influence of prolonged low protein feeding on epithelialized gingival wounds in rats, *J. Dent. Res.*, 45, 1448, 1966.
15. Bavetta, L. A., Bernick, S., Geiger, E., and Bergren, W., The effect of tryptophane deficiency on the jaws of rats, *J. Dent. Res.*, 33, 309, 1954.
16. Bavetta, L. A. and Bernick, S., Effect of tryptophan deficiency on bones and teeth of rats. II. Effect of prolongation, *Oral Surg. Oral Med. Oral Pathol.*, 9, 308, 1956.
17. Bavetta, L. A., Bernick, S., and Ershoff, B. H., Effect of tryptophan deficiency during lactation on bones and teeth of rats, *J. Dent. Res.*, 41, 366, 1962.
18. Bavetta, L. A., Bernick, S., and Ershoff, B., The influence of dietary thyroid on the bones and periodontium of rats on total and partial tryptophan deficiencies, *J. Dent. Res.*, 36, 13, 1957.
19. Bernick, S. and Bavetta, L. A., Histochemical and electron microscopy studies of dentin in tryptophan-deficient rats, *J. Dent. Res.*, 36, 142, 1957.

20. Bavetta, L. A. and Bernick, S., Lysine deficiency and dental structures, *J. Am. Dent. Assoc.,* 50, 427, 1955.
21. Messer, H. H. and Herold, R., Bone loss from different skeletal sites during acute calcium deficiency, *J. Dent. Res.,* 59, 902, 1980.
22. Ferguson, H. W. and Hartles, R. L., The effects of diets deficient in calcium or phosphorous in the presence and absence of supplements of vitamin D on the cementum and alveolar bone of young rats, *Arch. Oral Biol.,* 9, 647, 1964.
23. Ericsson, Y. and Ekberg, O., Dietetically provoked general and alveolar osteopenia in rats and its prevention or cure by calcium and fluoride, *J. Periodontal Res.,* 10, 256, 1975.
24. Coleman, R. D., Becks, H., Copp, D. H., and Frandsen, A. M., Skeletal changes of severe phosphorus deficiency in the rat. II. Skull, teeth and mandibular joint, *Oral Surg. Oral Med. Oral Pathol.,* 6, 756, 1953.
25. Weinmann, J. P. and Schour, I., Experimental studies in calcification. V. The effect of phosphate on the alveolar bone and the dental tissues of the rachitic rat, *Am. J. Pathol.,* 21, 1057, 1945.
26. Klein, H., Orent, E. R., and McCollum, E. V., The effects of magnesium deficiency on the teeth and their supporting structures in rats, *Am. J. Physiol.,* 112, 256, 1935.
27. Becks, H. and Furuta, W. J., Effect of magnesium deficient diets on oral and dental tissues. I. Changes in the enamel epithelium, *J. Am. Dent. Assoc.,* 26, 883, 1939.
28. Becks, H. and Furuta, W. J., Effect of magnesium deficient diets on oral and dental tissues. II. Changes in the enamel structure, *J. Am. Dent. Assoc.,* 28, 1083, 1941.
29. Becks, H. and Furuta, W. J., Effect of magnesium deficient diets on oral and dental tissues. III. Changes in the dentin and pulp tissue, *Am. J. Orthod. Oral Surg.,* 28, 1, 1942.
30. Gagnon, J., Schour, I., and Patras, M. C., Effect of magnesium deficiency on dentin apposition and eruption in incisors of rats, *Proc. Soc. Exp. Biol. Med.,* 49, 662, 1942.
31. Irving, J. T., The influence of diets low in magnesium upon the histological appearance of the incisor tooth of the rat, *J. Physiol.,* 99, 8, 1940.
32. Bernick, S. and Hungerford, F. G., Effect of dietary magnesium deficiency on the bones and teeth of rats, *J. Dent. Res.,* 44, 1317, 1965.
33. Barney, G. H., Macapinlac, M. P., Pearson, W. N., and Darby, W. J., Parakeratosis of the tongue — a unique histopathologic lesion in the zinc-deficient squirrel monkey, *J. Nutr.,* 93, 511, 1967.
34. Catalanotto, F. A. and Nandra, R., The effects of feeding a zinc-deficient diet on the taste acuity and tongue epithelium of rats, *J. Oral Pathol.,* 6, 211, 1977.
35. Meyer, J. and Alvares, O. F., Dry weight and size of cells in the buccal epithelium of zinc-deficient rats: a quantitative study, *Arch. Oral Biol.,* 19, 471, 1974.
36. Chen, S. Y., Cytochemical and autoradiographic study of nuclei in zinc-deficient rat buccal epithelium, *J. Dent. Res.,* 56, 1546, 1977.
37. Chen, S. Y., Ultrastructure of zinc-deficient rabbit oral mucosa, *J. Dent. Res.,* 59, 912, 1980.
38. Mesrobian, A. Z. and Shklar, G., The effect of dietary zinc sulphate on the healing of experimental extraction wounds, *Oral Surg. Oral Med. Oral Pathol.,* 28, 259, 1969.
39. Furstman, L. and Rothman, R., The effect of copper deficiency on the mandibular joint and alveolar bone of pigs, *J. Oral Pathol.,* 1, 249, 1972.
40. Monto, R. W., Rizek, R. A., and Fine, G., Observations on the exfoliative cytology and histology of the oral mucous membranes in iron deficiency, *Oral Surg. Oral Med. Oral Pathol.,* 14, 965, 1961.
41. Follis, R. H., Jr., Further studies on iodine-induced sialadenitis in hamsters and other species, *Lab. Invest.,* 12, 586, 1963.
42. Weatherall, J. A. and Weidmann, S. M., The skeletal changes of chronic experimental fluorosis, *J. Pathol. Bacteriol.,* 78, 223, 1959.
43. Schour, I. and Smith, M. C., Mottled teeth: an experimental and histologic analysis, *J. Am. Dent. Assoc.,* 22, 796, 1935.
44. Levy, B. M., Dreizen, S., Bernick, S., and Hampton, J. K., Jr., Studies on the biology of the periodontium of marmosets. IX. Effect of parathyroid hormone on the alveolar bone of marmosets pretreated with fluoridated and nonfluoridated drinking water, *J. Dent. Res.,* 49, 816, 1970.
45. Burn, C. G., Orten, A. U., and Smith, A. H., Changes in the structure of the developing tooth in rats maintained on a diet deficient in vitamin A, *Yale J. Biol. Med.,* 13, 817, 1941.
46. Schour, I., Hoffman, F. M., and Smith, M. C., Changes in the incisor teeth of albino rats with vitamin A deficiency and the effects of replacement therapy, *Am. J. Pathol.,* 17, 529, 1942.
47. Wolbach, S. B. and Howe, P. R., The incisor teeth of albino rats and guinea pigs in vitamin A deficiency and repair, *Am. J. Pathol.,* 9, 275, 1953.
48. Moore, T. and Mitchell, R. L., Dental depigmentation and lowered content of iron in the incisor teeth of rats deficient in vitamin A or E, *Br. J. Nutr.,* 9, 174, 1955.
49. Trowbridge, H. O., Salivary gland changes in vitamin A deficient rats, *Arch. Oral Biol.,* 14, 891, 1969.

50. Salley, J. J. and Bryson, W. F., Vitamin A deficiency in the hamster, *J. Dent. Res.*, 36, 935, 1957.
51. Dreizen, S., Levy, B. M., and Bernick, S., Studies on the biology of the periodontium of marmosets. XI. Histopathologic manifestations of spontaneous and induced vitamin A deficiency in the oral structures of adult marmosets, *J. Dent. Res.*, 52, 803, 1973.
52. Hayes, K. C., McCombs, H. L., and Faherty, T. P., The fine structure of vitamin A deficiency. I. Parotid duct metaplasia, *Lab. Invest.*, 22, 81, 1970.
53. Hirschi, R. G., Postextraction healing in vitamin A deficient hamsters, *J. Oral Surg.*, 8, 3, 1950.
54. Frandsen, A. M. and Becks, H., The effect of hypovitaminosis A on bone healing and endochondral ossification in rats, *Oral Surg. Oral Med. Oral Pathol.*, 15, 474, 1962.
55. Hale, F., Pigs born without eyeballs, *J. Hered.*, 24, 105, 1933.
56. Cohlan, S. Q., Congenital anomalies in the rat produced by excessive intake of vitamin A during pregnancy, *Pediatrics*, 13, 556, 1954.
57. Kalter, H., The teratogenic effect of hypervitaminosis A upon the face and mouth of inbred mice, *Ann. N.Y. Acad. Sci.*, 85, 42, 1960.
58. Deuschle, F. M., Geiger, J. F., and Warkany, J., Analysis of an anomalous oculodentofacial pattern in newborn rats produced by maternal hypervitaminosis A, *J. Dent. Res.*, 38, 149, 1959.
59. Abramovich, A., Changes in facial structures of rats caused by supplements of vitamin A and *Lathyrus odoratus*, *J. Dent. Res.*, 52, 300, 1973.
60. Cavalaris, C. J. and Krikos, G. A., Vitamin A produced mucous metaplasia, *J. Oral Ther. Pharmacol.*, 3, 452, 1967.
61. Lopes, R. A., Piccolo, A. M., Petenusci, S. O., daCosta, J. R. V., and Maia Campos, G., Effect of hypervitaminosis A on tongue muscles of the rat. A morphometric study, *Int. J. Vitam. Nutr. Res.*, 49, 235, 1979.
62. Lopes, R. A., Azoubel, R., Valeri, V., Iucif, S., and Gosuen, L. C., Morphologic effects of hypervitaminosis A on rat sublingual glands, *J. Dent. Res.*, 53, 757, 1974.
63. Gorlin, R. J. and Chaudhry, A. P., The effect of hypervitaminosis A upon the incisor and molar teeth, the alveolar bone, and temperomandibular joint of weanling rats, *J. Dent. Res.*, 38, 1008, 1959.
64. Dinnerman, M., Vitamin A deficiency in unerupted teeth of infants, *Oral Surg. Oral Med. Oral Pathol.*, 4, 1024, 1951.
65. Silverman, S., Jr., Eisenberg, E., and Renstrup, G., A study of the effects of high doses of vitamin A on oral leukoplakia (hyperkeratosis) including toxicity, liver function and skeletal metabolism, *J. Oral Ther. Pharmacol.*, 2, 9, 1965.
66. Abrams, E. J., Effects of thiamine deficiency on the nerves of the tongue of the rat, *J. Oral Ther. Pharmacol.*, 1, 406, 1965.
67. Perri, V., Rapuzzi, G., and Sacchi, O., Action of some thiamine antagonists on frog taste receptors, *Q. J. Exp. Physiol.*, 53, 381, 1968.
68. Afonsky, D., Oral lesions in niacin, riboflavin, pyridoxine, folic acid, and pantothenic acid deficiencies in adult dogs, *Oral Surg. Oral Med. Oral Pathol.*, 8, 315, 1955.
69. Levy, B. M., The effect of riboflavin deficiency on the growth of the mandibular condyle of mice, *Oral Surg. Oral Med. Oral Pathol.*, 2, 89, 1949.
70. Waisman, H. A., Production of riboflavin deficiency in the monkey, *Proc. Soc. Exp. Biol. Med.*, 55, 69, 1944.
71. Warkany, J. and Schraffenberger, E., Congenital malformations induced in rats by maternal nutrition deficiency. V. Effects of a purified diet lacking riboflavin, *Proc. Soc. Exp. Biol. Med.*, 54, 92, 1943.
72. Warkany, J. and Deuschle, F. M., Congenital malformation induced in rats by maternal riboflavin deficiency: dentofacial change, *J. Am. Dent. Assoc.*, 51, 139, 1955.
73. Deuschle, F. M., Takacs, E., and Warkany, J., Postnatal dentofacial changes induced in rats by prenatal riboflavin deficiency, *J. Dent. Res.*, 40, 366, 1961.
74. Sebrell, W. H. and Butler, R. E., Riboflavin deficiency in man (ariboflavinosis), *Public Health Rep.*, 54, 2121, 1939.
75. Afonsky, D., Oral lesions in niacin, riboflavin, pyridoxine, folic acid, and pantothenic acid deficiencies in adult dogs, *Oral Surg. Oral Med. Oral Pathol.*, 8, 206, 1955.
76. Afonsky, D., Experimental animal studies of tongue changes in nutritional disease, *Ann. N.Y. Acad. Sci.*, 85, 362, 1960.
77. Chittendon, R. H. and Underhill, F. P., The production in dogs of a pathological condition which closely resembles human pellagra, *Am. J. Physiol.*, 44, 13, 1917.
78. Goldberger, J. and Wheeler, G. A., Experimental black tongue of dogs and its relationship to pellagra, *Public Health Rep.*, 43, 172, 1928.
79. Smith, D. T., Persons, E. L., and Harvey, H. L., On the identity of the Goldberger and Underhill types of canine black tongue. Secondary fuso-spirochetal infection in each, *J. Nutr.*, 14, 373, 1937.

80. Denton, J. A., Study of the tissue changes in experimental black tongue of dogs compared with similar changes in pellagra, *Am. J. Pathol.*, 4, 341, 1928.
81. Elvehjem, C. A., Madden, R. J., Strong, F. M., and Woolley, D. W., The isolation and identification of the anti-black tongue factor, *J. Biol. Chem.*, 123, 137, 1938.
82. Sebrell, W. H., Onstott, R. H., Fraser, H. F., and Daft, F. S., Nicotinic acid in the prevention of black tongue of dogs, *J. Nutr.*, 16, 355, 1938.
83. Heath, M. K., MacQueen, J. W., and Spies, T. D., Feline pellagra, *Science*, 92, 514, 1940.
84. Belavady, B., Madhavan, T. V., and Gopalan, C., Production of nicotinic acid deficiency (black tongue) in pups fed diets supplemented with leucine, *Gastroenterology*, 53, 749, 1967.
85. Belavady, B. and Gopalan, C., Production of black tongue in dogs by feeding diets containing Jowar *(Sorghum vulgare)*, *Lancet*, 2, 1220, 1965.
86. Dreizen, S., Levy, B. M., and Bernick, S., Studies on the biology of the periodontium of marmosets. XIII. Histopathology of niacin deficiency stomatitis in the marmoset, *J. Periodontol.*, 48, 452, 1977.
87. Spies, T. D., Cooper, C., and Blankenhorn, M. A., The use of nicotinic acid in the treatment of pellagra, *JAMA*, 110, 622, 1938.
88. Levy, B. M., The effect of pyridoxine deficiency on the jaws of mice, *J. Dent. Res.*, 29, 349, 1950.
89. Afonsky, D., Oral lesions in niacin, riboflavin, pyridoxine, folic acid, and pantothenic acid deficiencies in adult dogs, *Oral Surg. Oral Med. Oral Pathol.*, 8, 438, 1955.
90. Berdjis, C. C., Greenberg, L. D., Rinehart, J. F., and Fitzgerald, G., Oral and dental lesions in vitamin B_6 deficient rhesus monkeys, *Br. J. Exp. Pathol.*, 41, 198, 1960.
91. Peer, L. A., Bryan, W. H., Strean, L. P., Walker, J. C., Bernhard, W. G., and Peck, G. C., Induction of cleft palate in mice by cortisone and its reduction by vitamins, *J. Int. Coll. Surg.*, 30, 249, 1958.
92. Mueller, J. F. and Vilter, R. W., Pyridoxine deficiency in human beings induced with desoxypyridoxine, *J. Clin. Invest.*, 29, 193, 1950.
93. Afonsky, D., Oral lesions in niacin, riboflavin, pyridoxine, folic acid, and pantothenic acid deficiencies in adult dogs, *Oral Surg. Oral Med. Oral Pathol.*, 8, 543, 1955.
94. Dreizen, S., Levy, B. M., and Bernick, S., Studies on the biology of the periodontium of marmosets. VIII. The effect of folic acid deficiency on the marmoset oral mucosa, *J. Dent. Res.*, 49, 616, 1970.
95. Dreizen, S. and Levy, B. M., Histopathology of experimentally induced nutritional deficiency cheilosis in the marmoset *(Callithrix jacchus)*, *Arch. Oral Biol.*, 14, 577, 1969.
96. Shklar, G., Effect of 4-amino-N^{10}-methyl-pteroylglumatic acid on the oral mucosa of experimental animals, *J. Oral Ther. Pharmacol.*, 4, 374, 1968.
97. Nelson, M. M., Asling, C. W., and Evans, H. M., Production of multiple congenital abnormalities in young by maternal pteroylglutamic acid deficiency during gestation, *J. Nutr.*, 48, 61, 1952.
98. Vogel, R. I. and Deasy, M. J., The effect of folic acid on experimentally produced gingivitis, *J. Prev. Dent.*, 5, 30, 1978.
99. Rose, J. A., Folic acid deficiency as a cause of angular cheilosis, *Lancet*, 2, 453, 1971.
100. Afonsky, D., Oral lesions in niacin, riboflavin, pyridoxine, folic acid, and pantothenic acid deficiencies in adult dogs, *Oral Surg. Oral Med. Oral Pathol.*, 8, 656, 1955.
101. Wainwright, W. W. and Nelson, M. M., Changes in the oral mucosa accompanying acute pantothenic acid deficiency in young rats, *Am. J. Orthod. Oral Surg.*, 31, 406, 1945.
102. Ziskin, D. E., Stein, G., Gross, P., and Runne, E., Oral, gingival and periodontal pathology induced in rats on a low pantothenic acid diet by toxic doses of zinc carbonate, *Am. J. Orthod. Oral Surg.*, 33, 407, 1947.
103. Levy, B. M., Effects of pantothenic acid deficiency on the mandibular joints and periodontal structures of mice, *J. Am. Dent. Assoc.*, 38, 215, 1949.
104. Frandsen, A. M., Becks, H., Nelson, M. M., and Evans, H. M., Growth and transformation of the mandibular joint in the rat. V. The effect of pantothenic acid deficiency from birth, *Oral Surg. Oral Med. Oral Pathol.*, 6, 892, 1953.
105. Nelson, M. M., Wright, H. V., Baird, C. D. C., and Evans, H. M., Teratogenic effects of pantothenic acid deficiency in the rat, *J. Nutr.*, 62, 395, 1957.
106. Ziskin, D. E., Karshan, M., and Dragiff, D. A., Oral manifestations in rats fed synthetic diets deficient in pantothenic acid and biotin, *J. Dent. Res.*, 27, 68, 1948.
107. Collins, D. A., Becks, H., Nelson, M., and Evans, H. M., Changes in the temperomandibular condyle of the rat resulting from nutritional disturbances including deficiencies in protein, phosphorus, vitamin A, riboflavin, biotin and pantothenic acid, *J. Dent. Res.*, 27, 741, 1948.
108. Sydenstricker, V. P., Singal, S. A., Briggs, A. P., DeVaughn, N. M., and Isbell, H., Observations on the "Egg White Injury" in man and its cure with a biotin concentrate, *JAMA*, 118, 1199, 1942.
109. Baugh, C. M., Malone, J. H., and Butterworth, C. E. Jr., Human biotin deficiency induced by raw egg consumption in a cirrhotic patient, *Am. J. Clin. Nutr.*, 21, 173, 1968.
110. Farrant, P. C., Nuclear changes in squamous cells from buccal mucosa in pernicious anemia, *Br. Med. J.*, 1, 1694, 1960.

111. Stone, R. E. and Spies, T. D., The effect of liver extract and vitamin B_{12} on the mucous membrane lesions of macrocytic anemia, *J. Lab. Clin. Med.*, 33, 1019, 1948.

112. Gorlin, R. J. and Levy, B. M., Changes in the mandibular joint and periodontium of vitamin B complex deficient rats and the course of repair, *J. Dent. Res.*, 30, 337, 1951.

113. Chapman, O. D. and Harris, A. E., Oral lesions associated with dietary deficiencies in monkeys, *J. Infect. Dis.*, 69, 7, 1941.

114. Glickman, I., Acute vitamin C deficiency in periodontal diseases. I. The periodontal tissues of the guinea pig in acute vitamin C deficiency. *J. Dent. Res.*, 27, 9, 1948.

115. Glickman, I., Acute vitamin C deficiency and the periodontal tissues. II. The effect of acute vitamin C deficiency upon the response of the periodontal tissues of the guinea pig to artificially induced inflammation, *J. Dent. Res.*, 27, 201, 1948.

116. Boyle, P. E., Bessey, O., and Wolbach, S. B., Experimental production of the diffuse alveolar bone atrophy type of periodontal disease by diets deficient in ascorbic acid (vitamin C), *J. Am. Dent. Assoc. Dent. Cosmos*, 24, 1768, 1937.

117. Dreizen, S., Levy, B. M., and Bernick, S., Studies on the biology of the periodontium of marmosets. VII. The effect of vitamin C deficiency on the marmoset periodontium, *J. Periodontal Res.*, 4, 274, 1969.

118. Hunt, A. M. and Paynter, K. J., The effects of ascorbic acid deficiency on the teeth and periodontal tissues of guinea pigs, *J. Dent. Res.*, 38, 232, 1959.

119. Boyle, P. E., Wolbach, S. B., and Bessey, O. A., Histopathology of teeth of guinea pigs in acute and chronic vitamin C deficiency, *J. Dent. Res.*, 15, 331, 1936.

120. Kalnins, V., Developmental disturbances of enamel in scurvy, *J. Dent. Res.*, 31, 440, 1952.

121. Avery, J. K., Han, S. S., and Lee, Y., Modifications of the fine structure of the incisor pulp of the guinea pig during experimental scurvy, *J. Dent. Res.*, 45, 440, 1966.

122. Levy, B. M. and Gorlin, R. J., The temperomandibular joint in vitamin C deficiency, *J. Dent. Res.*, 32, 622, 1953.

123. Irving, J. T. and Durkin, J. F., A comparison of the changes in the mandibular condyle with those in the upper tibial epiphysis during the onset of healing of scurvy, *Arch. Oral Biol.*, 10, 179, 1965.

124. Durkin, J. F., Irving, J. T., and Heeley, J. D., A comparison of circulatory and calcification changes induced in the mandibular condyle, tibial epiphyseal and articular cartilages of the guinea pig by the onset and healing of scurvy, *Arch. Oral Biol.*, 14, 1373, 1969.

125. Turesky, S. S. and Glickman, I. G., Histochemical evaluation of gingival healing in experimental animals on adequate and vitamin C deficient diets, *J. Dent. Res.*, 33, 273, 1954.

126. Kalnins, V., The effect of X-ray irradiation upon the mandibles of guinea pigs treated with large and small doses of ascorbic acid, *J. Dent. Res.*, 32, 177, 1953.

127. Hodges, R. E., Baker, E. M., Hood, J., Sauberlich, H. E., and March, S. C., Experimental scurvy in man, *Am. J. Clin. Nutr.*, 22, 535, 1969.

128. Weinmann, J. P. and Schour, I., Experimental studies in calcification. I. The effect of a rachitogenic diet on the dental tissues of the white rat, *Am. J. Pathol.*, 21, 821, 1945.

129. Weinmann, J. P. and Schour, I., Experimental studies in calcification. II. The effect of a rachitogenic diet on the alveolar bone of the white rat, *Am. J. Pathol.*, 21, 833, 1945.

130. Weinmann, J. P. and Schour, I., Experimental studies in calcification. III. The effect of parathyroid hormone on the alveolar bone and teeth of the normal and rachitic rat, *Am. J. Pathol.*, 21, 857, 1945.

131. Weinmann, J. P. and Schour, I., Experimental studies in calcification. IV. The effect of irradiated ergosterol and of starvation on the dentin of the rachitic rat, *Am. J. Pathol.*, 21, 1047, 1945.

132. Weinmann, J. P., Rachitic changes of the mandibular condyle of the rat, *J. Dent. Res.*, 25, 509, 1946.

133. Durkin, J. F., Heeley, J. D., and Irving, J. T., Comparison and changes in the articular, mandibular condylar and growth-plate cartilages during the onset and healing of rickets in rats, *Arch. Oral Biol.*, 16, 689, 1971.

134. Dreizen, S., Levy, B. M., Bernick, S., Hampton, J. K., Jr., and Kraintz, L., Studies on the biology of the periodontium of marmosets. III. Periodontal bone changes in marmosets with osteomalacia and hyperparathyroidism, *Isr. J. Med. Sci.*, 5, 731, 1967.

135. Fahmy, H., Rogers, W. R., Mitchell, D. P., and Brewer, H. E., Effects of hypervitaminosis D on the periodontium of the hamster, *J. Dent. Res.*, 40, 870, 1961.

136. Granados, H., Mason, K. E., and Dam, H., Dental changes of rats and hamsters in vitamin E deficiency, *J. Dent. Res.*, 25, 179, 1946.

137. Pindborg, J. J., The effect of vitamin E deficiency on the rat incisor, *J. Dent. Res.*, 31, 805, 1952.

138. Nelson, M. A., Jr., and Chaudry, A. P., Effects of tocopherol (vitamin E) deficient diet on some oral, para-oral and hematopoietic tissues of the rat, *J. Dent. Res.*, 45, 1072, 1966.

139. Tonge, C. H. and McCance, R. A., Severe undernutrition in growing and adult animals, *Br. J. Nutr.*, 19, 361, 1965.

140. Tamarin, A., Wanamaker, B., and Sreebny, L. M., The effect of inanition on the submandibular salivary glands and the exocrine pancreas of the rat, *Ann. N.Y. Acad. Sci.*, 106, 609, 1963.

141. Dreizen, S., Levy, B. M., and Bernick, S., Diet induced jejunal lipodystrophy in the cotton top marmoset *(Saguinus oedipus)*, *Proc. Soc. Exp. Biol. Med.*, 138, 7, 1971.

142. Dreizen, S., Levy, B. M., and Bernick, S., Studies on the biology of the periodontium of marmosets. XII. The effect of an experimentally produced malabsorption syndrome on the marmoset periodontium, *J. Periodontal Res.*, 7, 251, 1972.

143. Dreizen, S., Vogel, J. J., and Levy, B. M., The effect of experimentally induced atherosclerosis on the oral structures of the rabbit, *Arch. Oral Biol.*, 16, 43, 1971.

144. Dreizen, S., Stern, M. H., and Levy, B. M., Diet-induced arteriopathies in the rabbit aorta and oral vasculature, *J. Dent. Res.*, 57, 412, 1978.

145. Wirthlin, M. R., Jr., Greenberg, L. B., Hansen, L. S., and Ratcliff, P. A., Experimental atheroslcerosis: effect on arteries and periodontium of the rabbit mandible, *J. Periodontol.*, 42, 174, 1971.

146. Dreizen, S., Levy, B. M., and Bernick, S., Diet-induced atheroslcerosis in the marmoset, *Proc. Soc. Exp. Biol. Med.*, 143, 1218, 1973.

147. Doyle, J. L., Hollander, W., Goldman, H. M., and Ruben, M. P., Experimental atherosclorosis and the periodontium, *J. Periodontol. Periodontics*, 40, 350, 1969.

148. Plenk, H., Jr., Rudas, B., and Waechter, R., Evidence of alveolar bone loss in rats fed atherogenic diet, *J. Periodontal Res.*, 8, 106, 1973.

149. Krikos, G. A., Morris, A. L., Hammond, W. S., and McClure, H. H., Jr., Oral changes in experimental lathyrism (odoratism), *Oral Surg. Oral Med. Oral Pathol.*, 11, 309, 1958.

EXPERIMENTAL ORAL MEDICINE

Conservative estimates indicate that more than 200 different disease entities are expressed, sooner or later, by structural and functional alteration in the oral cavity. This pronounced reactivity of the mouth structures has provided a fertile field in which to study tissue responses to microbial, chemical, physical, metabolic, genetic, and, as reviewed in the preceding chapter, nutritional insults under experimental conditions. Animal models have provided a means of eliciting details of the dynamics of many of the conditions that affect the human mouth. As demonstrated below, such studies have contributed significantly to a fuller comprehension of the pathogenesis and natural history of a wide range of local and systemic stomatologic diseases.

EXPERIMENTAL HORMONE IMBALANCES

Pituitary Studies

Ziskin and Blackberg[1] examined the effects of hypophysectomy on the gingiva and oral mucous membranes of five female rhesus monkeys. They were sacrificed 16 to 60 days after undergoing complete surgical removal of the hypophysis. The heads were preserved in formalin for several weeks before removal of the gingiva and oral mucosa, which were then fixed in Bouin's solution, cleared, paraffin embedded, sectioned, and stained with hematoxylin and eosin, and with Masson's trichrome. In four of the five monkeys, most of the superficial epithelial layers were missing. Only the basal cell layer and a few layers of prickle cells remained. Hypophysectomy produced degenerative changes, faint staining properties, altered arrangement of the prickle cell layer, and intraepithelial separation in the gingiva and oral mucosa.

Dickler et al.[2] studied labeled cell doubling time of oral epithelium of the mouse interdental papillae in hypophysectomized animals. Twenty-five white mice, 45 to 50 days of age, with an average weight of 30 g, were subjected to pituitary gland removal. Three days later each received an i.p. injection of 1 μCi of tritiated thymidine per gram of body weight. Animals were sacrificed at intervals spaced from 30 min to 5 days. The mandibular molar area was dissected out, fixed in 10% formalin, demineralized, and embedded in paraffin. Mesiodistal sections were cut through the molar interdental papillae and each sixth section was deparaffinized and dipped into NTB$_3$ emulsion for radioautography. Slides were stained with nuclear fast red and indigo carmine. Six sections representing each time interval were examined microscopically. Mean labeled cell population, percent of labeled cells, and logarithms of the percent values were calculated. Doubling time of the labeled cell population was 16 hr in the hypophysectomized mice and 10 hr in normal mice. There was also slowed cell migration and cell proliferation time. Despite the upset in hormonal balance, the basic functions of mitosis and of migration of the oral epithelial cells continued to operate.

Carbone et al.[3] investigated the effects of hypophysectomy on the size of the submandibular gland and salivary proteins in weanling rats. Albino rats of the Charles River strain were divided by sex and litter at 21 days into a control group (22 animals) and experimental group (28 animals). Test animals were hypophysectomized on day 21. All rats were provided with water and a purified ration *ad libitum* and housed individually. At 60 days of age, whole saliva was obtained by pilocarpine stimulation and blood by cardiac puncture. Saliva was tested for total protein and serum, and then subjected to paper electrophoresis for protein partition. Hypophysectomy in the weanling rat resulted in major reductions in growth rate, food consumption, and in absolute and relative weights of the endocrine target organs. Pilocarpine-stimulated secretion and total salivary protein were closely related to body weight and to submandibular gland weight in both the normal and hypophysectomized rats. There was a notable increase in the anodal migration of salivary protein in the hypophysectomized animals.

Hypophysectomy-induced changes in the histochemistry of the rat submandibular gland were reported by Kronman and Chauncey.[4] In the group, 20 intact and 28 hypophysectomized 117-day-old Sprague-Dawley rats were sacrificed after a 30-day experimental period. Submandibular glands were removed, fixed in 10% formalin, processed, and sectioned. Slides were stained with hematoxylin and eosin, aqueous toluidine blue with RNA-ase control for RNA and metochromasia, PAS for localization of glycogen, neutral polysaccharides and water soluble glycoproteins, Alcian blue for acid mucopolysaccharides and histochemically for tryptophan and tyrosine as an index of amino acid and protein metabolism. Hypophysectomy caused pronounced atrophic changes in the submandibular gland. Protein metabolism was severely altered after pituitary removal, whereas acid mucopolysaccharides were unaffected. The tubular component of the submandibular gland was more sensitive to hypophysectomy than the acinar cells.

Hypophysectomy-related changes in the histochemistry of the rat submaxillary gland were studied by Shklar and Chauncey.[5] In this group, 17 hypophysectomized and 10 intact 117-day-old Sprague-Dawley rats were maintained on standard laboratory chow for 30 days. They were then sacrificed and the submaxillary-sublingual gland complex removed and split in half. One half was fixed in formalin and processed for histologic staining with hematoxylin and eosin. The other half was frozen immediately and stored at $-10°C$. As needed, the frozen halves were cut into blocks, thawed, fixed in chloral hydrate-formaldehyde, washed, embedded in gelatin, and cut with a freezing microtome. Sections were stained for acid phosphatases, nonspecific esterases, β-d-galactosidase, and succinic dehydrogenase by appropriate methods.

Morphologic and/or histochemical changes were noted in the acini, secretory tubules, and ducts consisting of reduction in cytoplasmic contents, disruption of cell architecture, and decrease in cell size. There was a diminution in acid phosphatase and β-d-galactosidase activity in the acini and ducts in contrast to increased activity in the secretory tubules. Succinic dehydrogenase activity in the secretory tubules and intralobular ducts was reduced. The evidence indicates that loss of pituitary function alters the enzymology in all three major cytologic components of the submaxillary gland.

The influence of testosterone, thyroxine, and cortisone on the submandibular and parotid glands of hypophysectomized rats was reported by Bixler et al.[6] Forty-eight albino rats, hypophysectomized at 40 days of age, were divided into four groups. Group I received daily s.c. injections of 10 μg sodium thyroxine in aqueous solution and 250 μg testosterone in sesame oil. Group II was given daily i.m. injections of 2.5 mg cortisone acetate in saline suspension. Group III was injected daily with 10 μg thyroxine, 250 μg testosterone, and 2.5 mg cortisone. Group IV was not injected with any hormones and served as controls. All animals were fed a special diet of natural foods *ad libitum* for 21 days, when they were sacrificed by ether inhalation. The salivary glands were fixed in 10% neutral formalin, embedded in paraffin, and sectioned and stained with Mallory's connective tissue stain and with hematoxylin and eosin.

Hypophysectomy produced an atrophy of the duct systems of the submandibular and parotid glands. Simultaneous administration of testosterone and thyroxine was effective in preventing hypophysectomy-induced changes in both glands. Cortisone per se appeared to have no significant effect on the histology of the rat salivary glands. When administered with testosterone and thyroxine, cortisone had a slight inhibitory effect on intralobular duct size in the submandibular gland and on the appearance of secretory granules in the submandibular and parotid glands.

Whether desiccated thyroid and/or testosterone prevents the regressive histologic and enzyme changes in the rat submaxillary gland that follows hypophysectomy was studied by Clark et al.[7] In the study, 110 female weanling Sprague-Dawley rats were hypophysectomized and 20 unoperated females were used as controls. The operated animals were distributed among four groups according to body weight. Group I received 5 to 10 mg desiccated thyroid daily in the diet. Group II was injected i.m. with testosterone in amounts ranging from 2.5 to 10 mg per week. Group III was given both thyroid and testosterone on the same schedule as Groups I and II. Group IV received neither thyroid nor testosterone. Three to four animals in each group were sacrificed at intervals ranging from one to six postoperative weeks. The salivary glands were divided for microscopic examination and analysis of proteolytic activity. Hypophysectomy prevented the morphologic development and suppressed the proteolytic activity of the rat submaxillary gland, changes not prevented by administration of desiccated thyroid or testosterone alone, or in combination.

Schour and van Dyke[8] described the changes in the incisor teeth of rats following hypophysectomy. The study was based on 22 successfully hypophysectomized white rats ranging in age from 36 to 64 days at the time of surgery. Intervals between operation and death were 63 to 459 days.

Grossly, hypophysectomy produced progressive retardation and ultimate cessation of eruption, disruption of form, and reduction in size of the incisor teeth. Histologically, the enamel epithelium underwent regressive changes and folding, resulting in either vicarious aborted enamel formation in the labial periosteum or in enamel hypoplasia. The ganoblasts became cuboidal and flat and finally underwent degeneration. Papillae became irregular and proliferated into the labial alveolar periosteum, where they became isolated in the form of eipthelial islands that subsequently degenerated. In animals living for more than 10 months postoperatively, enamel epithelium in the anterior zone was entirely lacking. In the other rats, the enamel in the anterior zone was usually normal except for occasional hypoplastic areas associated with connective tissue ingrowths. Enamel in the basal zone showed many variations from normal to defective structure and complete absence.

In the anterior zone, the pulp was almost completely obliterated and dentified. Residual pulp slit was mainly necrotic and contained many calcified globules. In the basal zone, the dentin showed multiple foldings and varied from normally calcified to very poorly calcified. In this zone, the pulp was typical in appearance, but the blood supply was reduced. Cementum was thicker than normal, sometimes increasing fivefold. Periodontal membrane was uneven in width, had a reduced blood supply, and contained an increased number of epithelial rests.

Ziskin et al.[9] determined the effect of hypophysectomy on the permanent dentition of rhesus monkeys. Six animals comprised the study group. In five, a hypophysectomy was successfully performed. The sixth monkey served as an operated control. All were given i.p. injections of alizarin red S at intervals before and after attempted hypophysectomy. After the animals were killed, ground sections were made of some of the teeth and measurements taken with an eyepiece micrometer. Hypophysectomy caused severe retardation of dentin apposition by affecting the calcification rhythm and gradient.

Thyroid Studies

Ziskin and Applebaum[10] studied the action of thyroidectomy and of thyroid stimulation on the growing permanent dentition of the rhesus monkey. The study group was comprised of ten females distributed as follows: five thyroidectomized, two castrated and thyroidectomized, three normal injected with a thyrotropic hormone. Each was given two or more injections of alizarin red S preexperimentally, followed by one to three injections i.p. during the study period of 10 and 100 days.

Thyroid ablation caused a marked retardation in the rate of dentin apposition. In the castrated-thyroidectomized animals, dentinal growth was almost completely impeded. In all of the thyroidectomized monkeys calcification of the enamel was abnormal, the predentin zone was greatly increased in width, and the postoperative dentin was so poorly calcified as to be undetectable by Grenz-ray. In those given thyrotropic hormone, there was an increase in rate of dentin apposition, a normal to slightly increased width of predentin, and normal dentinal calcification.

English[11] examined the effects of the goiterogenic agents, thiouracil and selenium, on the teeth and jaws of dogs. Four mongrel females were used in the experiment. Thiouracil was administered to the first dog at a level of 5 mg/kg/day; organic selenium was given the second and third dogs at doses of 5 mg/kg/day and 25 mg/kg/day, respectively. The fourth served as an untreated control. Drug administration was begun at weaning (6 weeks) and continued for 25 weeks. Animals were sacrificed and the jaws and teeth prepared for histologic study.

Lesions found in the animals given the goiterogens included enamel hypoplasia in the thiouracil-treated animal, structural defects presenting as interglobular dentin in the dog given the high dose of selenium, and thinning of the trabeculae in the mandibles of those given thiouracil and high-dose selenium. There was marked variation in the size and shape of the mandibular condyle in all drug-treated dogs with histologic indication of an untimely continuation of bone formation in the condyles of the selenium-dosed animals.

Goldman[12] induced hyperthyroidism in guinea pigs by feeding thyroid extract to determine the influence of this hypermetabolic state on the teeth and periodontium. In the study, 32 guinea pigs weighing between 265 and 420 g were given 0.6 mg thyroid extract by mouth every other day. The thyroid was made into a solution with water and administered orally through a medicine dropper. Animals were fed a stock diet plus a liberal supply of green vegetables. Survival time on the experimental regimen was from 6 to 104 days.

During the study period, the guinea pigs lost weight, developed thin and scraggly hair, and experienced nail degeneration and nail loss. All had generalized osteoporosis that was particularly evident in the femurs, tibias, and skulls. Other principal changes involved the pulp and dentin, where the ondontoblasts became deranged with pyknotic nuclei, and there was a decrease in the number of tubules and increase in tortuosity. Odontoblasts and blood vessels became enclaved in dentin, producing the appearance of osteodentin.

Clark et al.[7] elicited the effects of thyroxine on the structure and proteolytic activity of the rat submaxillary gland. In this study, 73 female Sprague-Dawley rats were divided into three groups. Group I was untreated controls; Group II had 10 to 20 mg desiccated thyroid per animal per day added to the diet; Group III was given 0.1% propylthiouracil by diet incorporation. Four or five animals from each group were killed between 3 and 15 study weeks. The submaxillary and thyroid glands from one side were fixed in Zenker's formalin and processed for histologic examination. The opposite freshly removed unfixed glands were used for determination of proteolytic activity. Desiccated thyroid caused an increase in the size and granular content of the serous tubules and an increase in the proteolytic activity of the submaxillary gland. In contrast, propylthiouracil suppressed both tubular and proteolytic activity.

The incidence of congenital malformations produced in laboratory animals by hypervitaminosis A or by X-irradiation can be modified by hormones or by hormone antagonists. Methylthiouracil augments the teratogenic activity of vitamin A whereas thyroxine protects against this effect of the vitamin. To determine whether the modifying action extends to deformities that are genetically determined rather than teratogenic in origin, Woollam and Millen[13] investigated the influence of thyroxine on the incidence of harelip in mice with a genetic propensity for this anomaly. Female mice of the Strong A line were used for brother-sister matings. Onset of pregnancy was determined by the appearance of a vaginal plug that denoted day 1. Animals were then divided into two groups. Group I constituted the untreated controls; Group II the test series in which the mice were given 0.1 mg thyroxine s.c. on days 11 and 12 of pregnancy. All were killed on day 18, the young removed from the uterus and examined for the presence of harelip. The control group of 207 young had 25 (12.17%) harelips, whereas the test group of 138 young had only 6 (4.4%) harelips, a difference that was highly significant statistically. The precise mechanism whereby thyroxine protected against the deleterious effects of the harelip gene or genes was not identified.

Hoskins and Asling[14] detailed the effects of thyroxine on endochondral osteogenesis in the mandibular condyle and in the proximal tibial epiphysis of rats. In this study, 20 Long-Evans rats were hypophysectomized at 26 to 28 days of age and divided into four groups of five each. One group served as a control, and the other three were given replacement therapy starting the day of the operation. Group I received 50 to 200 μg growth hormone per day, Group II animals were given 2.5 μg thyroxine per day, and Group III received both hormones daily in the aforementioned doses. Thyroxine augmented the growth hormone effect on endochondral osteogenesis in the mandibular condyles and in the proximal tibial epiphysis of the hypophysectomized animals. Graded growth hormone doses stimulated graded responses in the noncalcified region of the tibial epiphyseal cartilage plate.

Adrenal Studies

Schour and Rogoff[15] delineated the changes in the rat incisor following bilateral adrenalectomy. In this study, 45 young rats weighing between 27 and 95 g had both adrenals surgically removed. Nonoperated animals served as controls. All were fed a stock diet. Eight of the adrenalectomized rats died within 1 to 3 days; 30 died of acute adrenal insufficiency between 4 and 20 days. The remaining were sacrificed 43 to 44 days after surgery.

Characteristic changes were found in the incisors of all but one of the rats that survived up to 20 days, including those that lived only 1 to 3 days. Adrenalectomy-associated disturbances in the calcification of dentin were more common in the upper incisors than in the lowers. These included globules disseminated throughout the predentin in the middle third of the incisor, deep staining of the labial dentin by hematoxylin, and prominent stratifications in the lingual dentin. In animals surviving to 10 days postsurgery, postoperative dentin was distinguished from preoperative dentin by deeply stained bands that corresponded to dentin laid down at the time of adrenalectomy. In the seven animals whose prolonged survival was attributable to accessory adrenal bodies, the incisors underwent changes similar to those seen in rickets, notably, wide predentin and prominent interglobular dentin.

The cell kinetics of the oral epithelium of adrenalectomized mice were elicited by Dickler et al.[2] Twenty-five CD Number 1 white mice 45 to 50 days of age, with an average weight of 30 g were adrenalectomized. Three days later each was injected with 1 μCi of tritiated thymidine per gram of body weight. Single animals were sacrificed at intervals between 30 min and 5 days postinjection. Sample preparation for microscopic and radioautographic examinations was identical to that in previously described studies in hypophysectomized mice by the same investigators.[2] The doubling time of the labeled oral epithelial cell population in the adrenalectomized mice was 14 hr compared to 10 hr in normal mice. As in the hypophysectomized mice, cell migration and exfoliation patterns in the adrenalectomized animals were slower than normal.

Liu and Lin[16] studied the role of adrenocortical hormones in the growth and development of the salivary glands in immature rats. Weanling Sprague-Dawley rats were adrenalectomized or sham adrenalectomized, fed Purina® chow, and given 0.9%

NaCl solution for drinking *ad libitum*. Beginning on the second postoperative day, the adrenalectomized rats were injected s.c. with either 10, 40, or 320 μg corticosterone per 100 g body weight per day for 4 weeks. Sham-operated controls received the vehicle alone on the same schedule. At necropsy the salivary glands were removed and weighed. Submandibular gland sections were prepared and stained with Azan.

Adrenalectomy retarded both submandibular and total body growth evidenced by a decline in gland weight, granular tubule and striated duct diameters, and acinar cell size. With increasing doses of hormone, there was a proportionate increase in submandibular and total body growth. Submandibular and body weights of rats given 320 μg corticosterone were almost the same as in the sham-operated controls. Parotid gland growth was also retarded by adrenalectomy and restored to normal by 320 μg corticosterone.

Clinical and experimental evidence indicates that prolonged systemic administration of cortisone affects protein synthesis and bone growth. Since large doses of cortisone inhibit growth of the facial skeleton in immature rats, Che-Kuo and Johannessen[17] examined the effect of small doses of cortisone on the synchondroses of the cranial base that participate in the growth of the neural and facial craniums. In the study, 48 male albino rats, 21 days of age, were distributed into three groups. Group I received a diet containing 0.05 mg cortisone per gram of ground Purina® rat chow for 20 days. For Group II, 0.2 mg desiccated thyroid was added to each gram of cortisone diet. Group III was fed only Purina® rat chow throughout the experimental period. Each rat in Groups II and III was pair-fed with the corresponding animal in Group I. Postmortem radiographs were taken of the skulls and tibias after the rats were killed with chloroform and necropsied. Five long bone dimensions and ten craniofacial dimensions were measured on the radiographs. The tissues were fixed in 5% neutral formalin, washed in tap water, and stored in 50% ethyl alcohol. After routine histologic preparation, sections from the cranial base, tibia, temperomandibular joint, and thyroid and adrenal glands were examined.

Low doses of cortisone decidedly affected the growth of the tibia and of the anteroposterior dimensions of the cranial base. Desiccated thyroid did not counteract but may have augmented the effects of cortisone. Skulls of the cortisone-fed rats had greater intracranial height than the pair-fed controls. Histologically, the tibia, cranial base, and temperomandibular joint of the rats on the cortisone-containing diet showed retarded endochondral ossification and partial inhibition of remodeling and resorption of the primary spongiosa.

Continued cortisone overdosage produces osteoporosis of the alveolar bone in young mice and rats. Applicability of these findings to primates is tempered by an apparent species specificity in the osteopenic action of cortisone. The rat, unlike the rabbit and man, does not develop osteoporosis when given an overabundance of cortical steroids unless the calcium and phosphorus content of the diet is low or unbalanced. Although overproduction of the antianabolic glucocorticoids has been implicated as a cause of human osteoporosis, there was a paucity of experimental data relative to the influence of cortisone on adult alveolar bone until Dreizen et al.[18] determined the effect of cortisone on the periodontium and skeleton of adult marmosets.

Five male and five female healthy adult cotton top marmosets (*Saguinus oedipus*) were selected at random and given daily i.m. injections of 10 mg cortisone acetate. Five untreated adult animals — housed, fed, and watered in the same room — served as controls. One animal from the control group and two from the experimental series were killed by ether inhalation at the end of 30 study days and after two, three, four, and six study months. They were necropsied immediately and the tissues fixed in 10% neutral formalin. Following fixation, the heads were sawed midsagittally and Grenz-rayed along with the remainder of the skeleton. The skeletal specimens were decalcified, paraffin processed, sectioned and stained with hematoxylin and erythrosin B, Masson's trichrome, Mallory's connective tissue stain, PAS and hematoxylin, and Weigert's iron hematoxylin.

Histologically distinguishable differences between the cortisone-treated and untreated marmosets were most conspicuous in the alveolar bone, skull, axial and appendicular skeleton, adrenals, and kidneys. Scope and severity were time related. They followed a cumulative progression starting with a suppression of the inflammatory response and extending through calcification in the renal medulla, atrophy of the adrenal cortex, osteoporosis of the alveolar bone, vertebrae, and long bones, and diminished fibroplasia in the periodontal ligament. Each of the cortisone-treated animals had histologic evidence of a generalized loss of bone mass when contrasted with control animals. In the alveolar bone, this was reflected by an initial thinning and an eventual dissolution of the trabeculae and bony plates surrounding the teeth in the upper and lower jaws. The bone loss was not accompanied by excessive osteoclastic activity, osteoblastic abnormalities, production of increased amounts of uncalcified osteoid, or bone marrow fibrosis as seen in hyperparathyroidism. Instead, the bone loss was atrophic in type and manifested by a quantitative diminution in mass without any histologically detectable compositional alteration. Although the precise mechanism is unknown, the preponderance of animal and human data suggests that superfluous cortisone promotes an increase in bone resorption, a heightened return of calcium from the skeleton, and a probable decrease in bone production (Figure 1A, 1B, 1C).

Selye[19] produced a nomalike condition in rats while administering cortisone and growth hormone to test the resistance of the buccal mucosa. In the study, 32 female piebald rats weighing 120 to 135 g fed Purina® fox chow were divided into four groups. Group I was injected s.c. with 1 mg cortisone per day during week 1, and 2 mg/day thereafter. Group II received somatotrophic hormone s.c. at a level of 1.5 mg twice a day. Group III was given both hormones in the doses mentioned. Group IV was untreated and acted as controls. The lower incisors were cut down to the level of emergence from the gingiva on the first day of the experiment in all groups and retrimmed every 3 or 4 days during the period of observation. The rats were killed on day 29 and the mandibles removed for histologic examination.

Those given cortisone alone developed an extensive and progressive gangrenous stomatitis or noma. Ulcers began to appear on the gingival surface around the lower incisor roots after 20 days of treatment. In the next 5 days, every animal in this group had such ulcers. At first the gingival surface was only slightly eroded and covered with food debris mixed with necrotic tissue. Subsequently, destruction of the soft tissue proceeded along the anterior and ventral surface of the mandible, gradually spreading to the buccal mucosa.

Earlier studies had shown that under such conditions the destructive lesions gradually perforated the skin, causing extensive and disfiguring defects. In the present series, the animals were killed before this stage was reached to obtain material for histologic examination. Microscopically, there was a nomalike necrotizing ulcer on the inner surface of the lips that extended deep into the underlying tissue. Dead tissue was not separated from the deeper layers by a new connective tissue barrier. Clipping of the lower incisors and forcing the animals to chew with their gingiva exposed the buccal mucosa to excessive mechanical injury. Under such conditions, the ensuing gingival ulceration was presumably prevented from healing by cortisone, leading ultimately to the destructive stomatitis.

Gonadal Studies

The changes in the rodent incisor induced by bilateral gonadectomy were described by Schour.[20] In that study, 26 female ground squirrels (*Citellus tridecemlineatus*) were

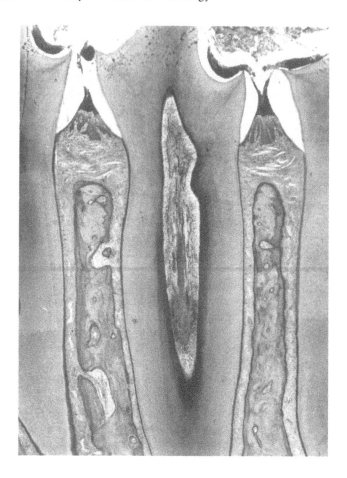

FIGURE 1A.   Mandibular alveolar bone in control animal. PAS and hematoxylin stain. (Magnification × 35.) (From Dreizen, S., Levy, B. M., and Bernick, S., *J. Periodontol.*, 42, 217, 1971. With permission.)

subjected to bilateral gonadectomy at between 2 and 14.5 months of age. The test animals were killed with ether from 25 to 339 days after surgery, together with assigned untreated controls. Heads were removed, fixed in 10% formalin, sawed midsagittally, and radiographed. Tooth-bearing areas were blocked, decalcified, dehydrated, embedded in celloidin, sectioned, and stained with hematoxylin and eosin or with iron hematoxylin and Mallory's connective tissue stain. Radiographs revealed an abnormal amount of radiolucent dentin in 54% of the experimental animals. Histologically, the organic enamel matrix persisted to an abnormal extent in the middle or distal third of the incisor. The dentin had three distinct zones of disturbed calcification that disregarded the incremental pattern. Severe cases also showed disruption of dentin formation. Alveolar bone manifested a compensatory hyperplasia of osteoid at the lingual alveolar crest.

Shafer and Muhler[21] investigated the influence of gonadectomy and of sex hormones on the structure of the rat salivary glands. A total of 134 albino rats, of which 59 males and 58 females completed the experiment, were divided into eight groups on the basis of sex, weight, and litter mate distribution. The males were grouped as Group I, control; Group II, control given testosterone; Group III, gonadectomized; Group IV, gonadectomized plus testosterone. The four female groups were treated identically except for receiving diethylstilbestrol instead of testosterone. Testosterone was adminis-

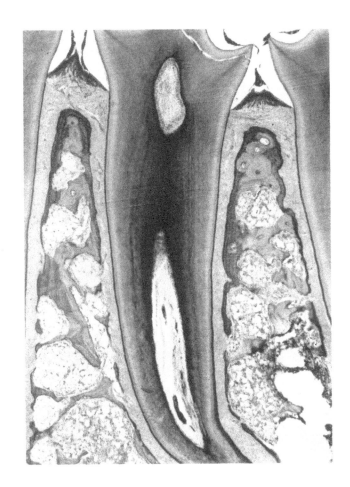

FIGURE 1B. Osteoporosis of alveolar bone in mandible of marmoset given cortisone for 3 months. (PAS and hematoxylin stain; Magnification × 35.) (From Dreizen, S., Levy, B. M., and Bernick, S., *J. Periodontol.*, 42, 217, 1971. With permission.)

tered by s.c. implantation of a pellet weighing 37 mg. After 45 days, a 75-mg pellet was similarly implanted. At the end of the experiment, the pellet residues were removed and weighed for calculation of total dose. Females were injected s.c. with diethylstilbestrol in oil at a concentration of 10 mg/week. All animals were 30 days of age at the beginning of the study and 147 at the end.

At termination, the salivary glands were removed en masse, fixed in Zenker's formalin, dehydrated, paraffin-embedded, and stained with hematoxylin and eosin, phosphotungstic acid-hematoxylin, or modified Mallory's connective tissue stain. The chief histologic changes were in the granular tubules of the submaxillary gland. Both male and female gonadectomized rats had a decrease in size and number of secretory tubules compared to controls. In males, testosterone increased the size and number of granular tubules. In females, diethylstilbestrol had an opposite effect, regardless of whether or not the animals were gonadectomized.

Ziskin and Blackberg[1] looked into effects of castration on the oral mucosa of four female and three male rhesus monkeys. Specimens from control untreated monkeys consisted of biopsies of interdental papillae and from the castrates of large sections of gingival and buccal mucosa. In the females, castration caused alterations in the prickle cell layer of the alveolar and attached gingiva and tissue degeneration. Castration of

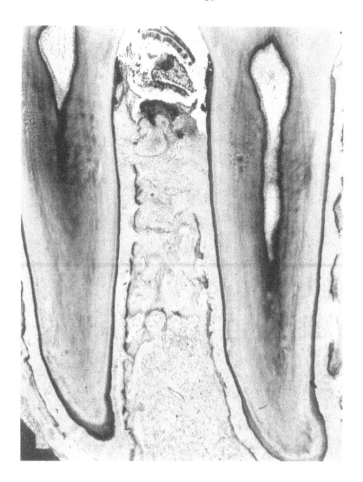

FIGURE 1C.   Extensive osteoporosis of alveolar bone in mandible of
marmoset treated with cortisone for 6 months. (PAS and hematoxylin
stain; Magnification × 35.) (From Dreizen, S., Levy, B. M., and Bernick,
S., *J. Periodontol.*, 42, 217. 1971. With permission.)

male monkeys induced hyperkeratinization of the gingival and oral mucosa similar to
that following injections of estrogenic hormone.

The effect of estrogens on the gingiva and alveolar bone of rats and mice was reported by Nutlay et al.[22] In this study, 16 newborn rats were divided into eight controls
and eight recipients of 0.05 to 1.30 mg α-estradiol benzoate by injection. All were
sacrificed 5 to 42 days later. Of a group of newborn mice, 22 were segregated into
eight controls; 6 were given 0.05 to 0.80 mg diethylstilbestrol by weekly injections and
sacrificed at 5 to 30 days of age, and 8 were given a total of 0.05 to 1.30 mg α-estradiol
benzoate and killed at 5 to 40 days of age. A group of 23 mice, 1 year of age, were
distributed among six controls; 9 received a s.c. pellet of about 2 mg crystalline α-
estradiol benzoate and sacrificed after 1 to 13 study weeks, and 8 were given a total
of 1.5 to 6.0 mg α-estradiol benzoate in 0.50 mg weekly injections and terminated 3
to 12 weeks later.

Whereas the alveolar bone and gingiva of the very young rats were unaffected by
administration of estrogens, the newborn mice treated with diethylstilbestrol or α-estradiol benzoate developed marrow fibrosis in the alveolar bone and a reduction in
marrow spaces by new bone formation. In the old mice subjected to s.c. pellets of α-
estradiol benzoate, there was a downgrowth of the epithelial attachment along the root

surface extending into the root bifurcation. The gingival epithelium of the interdental papillae between the first and second molars proliferated in an almost tumorlike fashion. Interdental papillae lost their conical shape and became progressively more blunted and concave. In the final stages, there were deep interdental and interradicular pockets that led to a marked resorption of the interdental and interradicular bony septa. Root surfaces had many areas of resorption.

The influence of progesterone on the growth and development of the submandibular glands and other glands in female rats was evaluated by Liu and Lin.[23] Weanling female Sprague-Dawley rats were given daily s.c. injections of 0.6 to 1.5 mg progesterone per 100 g body weight for 10 weeks; controls received diluent vehicle only in the same manner. Total body and gland weights were obtained at necropsy. The numbers and diameters of the granular tubules and the size of the acinar cells in the submandibular gland were measured in histologic sections. Progesterone did not cause significant changes in body or gland growth at the low dose except for a decrease in adrenal gland weight. At the high dose, progesterone prompted slight increase in submandibular gland and total body weight, a significant increase in number of granular tubules and acinar cell size in the submandibular gland, a significant decrease in thyroid gland weight, and no meaningful change in either parotid or adrenal gland weight in comparison to controls.

Pancreatic Studies

Kreshover et al.[24] determined the effect of diabetes on tooth development in the rat. The course of pregnancy was followed in 29 albino rats given alloxan at various times after mating at a mean i.p. dose of 155 mg/kg body weight. Protamine zinc insulin was given at a frequency and dose governed by the degree of hyperglycemia and glycosuria. In the event of a successful pregnancy, the young were sacrificed from 1 to 60 days of age. Sections of pancreas and jaws were prepared for histologic study.

Of 31 rats with positive evidence of pregnancy, 22 delivered spontaneously at term, 1 by Caesarean section, 2 probably had fetal resorption, 1 evidenced resorption or abortion, and 5 died before delivery. The later during pregnancy that alloxan was given, the greater the success of the pregnancy. All but one of the eight rats that did not become pregnant received alloxan prior to the 12th day after mating. In no instances were either excessive weight or elevated blood sugar found in the young of the alloxan-injected rats.

Almost all of the animals receiving alloxan showed abnormalities of amelogenesis. Most common was the presence of enamel-like material on the surface of the organic matrix. These varied in shape from small globular formations to large irregular amorphous deposits. In some instances, the overlying ameloblasts were fragmented and vacuolated. Under the degenerated ameloblasts was an abnormally formed organic enamel matrix characterized by pale-staining properties, irregularly arranged prisms, and dark-staining fiberlike formations. Most of the alloxan-injected rats showed vacuolization of the odontoblasts and pulpal hyperemia and hemorrhage. Whereas the changes in the enamel organ were confined to the incisors, pulpal lesions were also present in the molars. The nature and degree of these changes were not related either to time of drug administration or severity of the resulting hyperglycemia.

The diabetic Chinese hamster shows pathologic derangements similar to those in human diabetes and seems, therefore, to be an appropriate animal model for diabetic research. Diabetes in the Chinese hamster is primarily pancreatogenic, with the beta cells exhibiting degranulation, karyolysis, glycogen infiltration, and eventual disappearance. The disease is apparently genetically determined as a recessive factor. Anapolle et al.[25] used this model to study diabetes-related microvascular changes in gingiva and cheek pouch tissue. Gingival tissue was removed from the labial aspect of the incisors, from the cheek pouch of diabetic Chinese hamsters, and from normal controls by excisional biopsy. The samples were subjected to standard electron microscope fixation and preparation. All animals were matched pairs of similar age. Vessels were examined under 5000 and 15,000 times magnification. In the diabetic animals the small blood vessels in the gingiva, oral musculature, and cheek pouch had a marked thickening of the basement membrane and were surrounded by an increased amount of amorphous basement membranelike substance.

Diabetes mellitus accelerates periodontal breakdown and impairs wound healing in the oral cavity of man and experimental animals. That such action may result from enhanced collagenase activity in the gingiva is suggested by the studies of Golub et al.[26] Four-month-old male Sprague-Dawley rats weighing approximately 350 g were made diabetic by either i.p. injection of alloxan monohydrate (12 mg/100 g body weight) or by i.v. injection of streptozotocin (7 mg/100 g body weight) following a 4-day fast. The rats were killed with ether 21 days later, and the gingiva of the maxillary incisors was excised and assessed for collagenase activity in tissue culture.

Gingiva from both alloxan and streptozotocin diabetic rats showed significantly increased collagenolytic activity compared to the nondiabetic controls. Soluble extracts of the diabetic gingiva *in situ* contained breakdown products of collagen similar in size to reaction products generated by tissue collagenase. Such fragments were not detected in the control specimens. Diabetes appears to stimulate synthesis of gingival collagenase.

The role of insulin in total body, salivary gland, and endocrine gland growth in rats was elucidated by Liu and Lin.[27] Female Sprague-Dawley rats were used in three experiments. The rats were made diabetic at 22 or 53 days by s.c. injection of 20 mg alloxan per 100 g body weight after a 24-hr fast. Diabetes was determined by the presence of glycosuria, diuresis, and polydipsia. In the first experiment, rats rendered diabetic at 23 days of age were randomly divided into three groups. Two groups were given daily s.c. injections of 1 to 2 USP units insulin per 100 g body weight for 14 days. The third group of terminal diabetic and intact controls was treated only with the insulin vehicle. The second experiment was similar to the first, differing only in that 3 USP units were given each rat for 14 days. Experiment three examined the effect of alloxan diabetes and insulin replacement in mature rats made diabetic at 53 days of age.

At the completion of each experiment, the animals were killed and the salivary, thyroparathyroid, parathyroid, adrenal, and anterior pituitary glands immediately removed and weighed. Salivary glands were fixed in 10% formalin, and sections of the submandibular glands were stained with Azan. The diameters of the intralobular striated ducts and convoluted granular tubules were measured in the stained sections. Mean body weight, epiphyseal cartilage width of tibia, salivary and endocrine gland weights, mean diameter of striated ducts, or granular tubules of the diabetic rats were compared statistically with those of the controls.

Alloxan-induced diabetes in the immature rats resulted in retardation or cessation of body growth and growth of the salivary and endocrine glands. In the mature rats, alloxan diabetes resulted in a significant decrease in salivary gland and total body weight. The ductal and tubule diameters of the submandibular glands of the mature diabetic rats decreased significantly compared to controls. Insulin replacement therapy prevented the change in the salivary and endocrine glands in both immature and mature diabetic rats. With large doses of insulin, salivary gland, endocrine gland, and total body growth of the immature diabetic rats exceeded that of normal controls.

Enlargement and functional changes in the salivary glands are regularly found in patients with cystic fibrosis of the pancreas. By feeding pancreatin to rats, Mangos et al.[28] produced parotid and submaxillary gland enlargement and changes in the secretory activity, handling of electrolytes, and excretion of protein by these glands. Male Sprague-Dawley albino rats born on the same day were assigned to groups at random in each experiment. The experimental diet consisted of a mixture of ground rat ration and pancreatic enzymes in the form of either pancreatin powder, pancrelipase, or pancreatin granules. Each was added to the ground diet at a 4% level. Control animals received only the ground rat meal. Most of the experiments were performed after the rats had been on the diets for 3 weeks.

In the rats fed 4% pancreatic enzymes, the weights of the parotid and submaxillary glands increased 1.7 and 2.7 times, respectively. Enlargement in the submaxillary glands was due to true nonaqueous cellular material. There was a marked increase in the size of the acinar cells, with the cytoplasm developing a mucoid appearance. When rats were fed 4% pancreatic enzymes heated to 100°C for 3 hr, gland enlargement did not occur. Gastric intubation of the 4% pancreatin preparations had no effect on the size of the salivary glands. Rats fed 4% pancreatic enzymes and given s.c. injections of 5 mg theophylline twice daily had a notable increase in salivary gland size. Functional alterations in the salivary glands included (1) slightly greater slowing of parotid flow rate in response to i.v. pilocarpine (0.1 mg/100 gm body weight) in experimental than in control rats, and (2) significantly lower submaxillary flow rates in experimental animals in response to pilocarpine compared to controls. Submaxillary salivary sodium, potassium, calcium, and protein concentrations were higher in the test than in the control rats. In rats fed 4% pancreatic enzyme, the saliva of both glands exhibited prominent changes in the electrophoretic patterns of excretion of basic secondary protein. Sialadenotrophic and functional effects in the salivary glands of rats given the pancreatic enzymes were similar to those in the salivary glands of patients with cystic fibrosis of the pancreas.

EXPERIMENTAL ORAL INFECTIONS

Bacterial Infections

Kennedy et al.[29] studied effects of hemolytic streptococci inoculated into the root canals of the teeth of monkeys in an effort to unravel the relationships between focal infections, bacteremia, rheumatic disease, and specific alterations in streptococcal antibodies. In the first experiment, six monkeys were inoculated alternately with β-hemolytic streptococci, Groups A and C, smooth phase. Five monkeys served as uninoculated controls. Pulp was removed from the anterior teeth of all animals. Teeth in the test group were seeded with cultural sediment suspended in FALBA. A second experimental series of three monkeys was inoculated in the same fashion with Group A streptococci, matt phase. Whereas the smooth phase organisms failed to establish a bacteremia or to elicit any systemic reactions, the matt phase organisms produced bacteremia regularly along with fever, leucocytosis, and elevated erythrocyte sedimentation rates. Serologic response was evidenced by an increasing titer of antistreptolysin O. Histologically, each of the animals in this group had microscopic evidence of cardiovascular disease of the "rheumatic" type.

Okada et al.[30] developed an experimental model of dental focal infection in rabbits by prolonged sensitization with horse serum protein through the dental pulp canals. Test animals were adult rabbits weighing 1.5 to 3 kg. Lyophilized horse whole serum was used as the antigen. Under Nembutal® anesthesia, the right buccal region was exposed surgically to the anterior side of the masseter. The pulps of the right upper molars were extirpated and the root canals enlarged. Antigen in 5 to 20 mg dry weight per dose was mixed into a paste with physiological saline containing 1/10,000 Merthiolate® and inserted into the pulp canals that were sealed with eugenol cement. Antigen was reapplied at intervals of 7 to 10 days. Each rabbit was sensitized from 3 to about

30 times over a 1- to 10-month period. Immunopathologic studies were made on changes in serum proteins and antibodies and on local skin reactions (Arthus, passive cutaneous anaphylaxis, and Prausnitz-Kustner reactions) and systemic changes (rheumatoid factorlike substance and autoantibodies to organ extracts).

Changes in the serum protein fraction were characterized by hypoalbuminuria and hypergammaglobulinemia that peaked between 50 and 100 days after initial sensitization. The hypergammaglobulinemia was coincident with a high titer of specific precipitating antibody. Both precipitin and sensitized red cell agglutination titers increased progressively until the end of the experiments. Arthus reaction and passive cutaneous anaphylaxis reaction were positive. Locally, an allergic demarcating inflammation with a thick granulation tissue layer and vascular lesions including periarteritis and focal histolysis were demonstrated. Slight endocarditis, myocarditis, diffuse interstitial pneumonitis, various forms of hepatitis, diffuse glomerulonephritis, and splenitis were observed. In the submandibular lymph nodes, degenerative changes such as onion skin vascular lesions were seen during the later stages of sensitization. Repeated administration of the antigen through the pulp canals caused both a local and systemic immunologic response. The systemic morphologic changes were allergic in origin. The evidence suggests that an antigenic substance or substances in chronic dental foci might cause a continuous immunologic stimulus that induces systemic disease.

Allard et al.[31] inoculated pure cultures of well-characterized microorganisms into root canals of dogs to determine whether this would lead to anachoresis or the recovery of the test organisms from a lesion other than the site of inoculation. Fourteen dogs were subjected to partial and/or total pulpectomies on most premolars in each quadrant. These were done aseptically. Pulpectomized roots were inoculated with culture broth alone or culture broth of four different bacteria, notably *Staphylococcus aureus, Streptococcus sanguis, Pseudomonas aeruginosa* and *Bacteroides fragilis.* Only one bacterial species was used for each dog. The teeth were sealed with sterile cotton, fast-setting zinc oxide cement, and amalgam fillings. Bacteriologic samples were taken repeatedly between 4 and 120 days by means of a sterile charcoaled paper point wetted with sampling fluid and placed first into the canal and then into transport medium. Similar samples were obtained from the inoculated teeth after 50 and 100 days to determine whether the test organisms were still viable. Cultivation was carried out aerobically and anaerobically. Radiographs of the jaws were taken before, during, and at the end of the experiment when the animals were killed, necropsied, and the jaws prepared for histologic examination.

Inoculated microorganisms were isolated from the majority of noninfected roots after 28 to 120 days, demonstrating that transport of organisms or anachoresis had occurred. Radiolucencies were present around most of the roots on X-ray. Where no radiographic changes could be demonstrated, no bacterial growth was obtained. Histologically, there was a chronic inflammatory reaction around the roots where *P. aeruginosa* and *B. fragilis* had been inoculated or isolated. The inflammatory reaction around roots inoculated with *Staphylococcus aureus* or *Streptococcus sanguis* was dominated by polymorphonuclear leucocytes.

Silver et al.[32] measured the clearance rates of oral microorganisms following the production of experimental bacteremias in dogs. Aerobic and anaerobic microorganisms from man (*S. mitis; S. sanguis; Corynebacterium* sp.) and dog (*Staphylococcus aureus; Escherichia coli ; B. melaninogenicus)* were inoculated singly into 12 mongrel dogs by the i.v. route. Inocula were prepared by subculturing the organisms from aerobic and/or anaerobic blood agar plates into Trypticase® soy broth for 18 hr. *B. melaninogenicus* was incubated in Bacteroides broth for 48 hr. All were grown at 37°C. Cells were harvested by centrifugation and resuspended in fresh growth medium before use. Concentrations of 1.3×10^6 and 2.0×10^7 cells per milliliter were inoculated into the external jugular vein or a superficial hind leg vein of the anesthetized animals after first obtaining a base line blood sample.

At precisely timed intervals, blood specimens were drawn aseptically from a fore leg vein into Vacucontainer tubes for plating on blood agar. Plates were incubated at 37°C for 18 hr aerobically and the colonies counted. Those inoculated with *B. melaninogenicus* were incubated for 4 days in an anaerobic chamber before counting. Maximum numbers of all bacteria were recovered within 30 sec in most animals and never later than 1 min after inoculation. The circulation was usually clear of organisms within 10 min, and always within 20 min. Rates of clearance from the blood were not dependent on bacterial species or source.

Bahn et al.[33] determined whether human oral streptococci inoculated into the oral cavity of rabbits with prepared sterile cardiac vegetations would consistently produce endocarditis. Sterile cardiac vegetations were established in 141 New Zealand rabbits weighing 2 to 3 kg. A sterile polyethylene catheter filled with sterile saline solution was introduced through the superficially opened right carotid artery to the aortic valve area. The catheter was then tied into the artery and the superior end folded upon itself and tied. One week after catheterization, the animals were divided into five groups and treated as follows:

	Group	No. of rabbits	Dental manipulation	Route	Inoculum
				Bacterial inoculum	
I					
	A	15	None	None	None
	B	4	Extraction of incisor	None	None
	C	6	None	i.v.	0.5 ml of 10^7 *S. mitis*
II					
	A	11	Extraction of incisor	Into bleeding extraction wound	0.1 ml of 10^7 *S. mutans*
	B	12	Wounds of tongue and gingiva	On sutures into wound	1.0 µl of 10^7 *S. mutans*
III					
	A	7	Gingiva wound	Into gingiva	1 ml of 10^7 *S. mitis*
	B	8	Extraction of incisor	Paper point into extraction wound	0.01 ml of 10^7 *S. mitis*
IV					
	A	12	Palatal wound	Into palate	1 ml of 10^7 *S. mutans*
	B	16	Palatal wound	Into palate	1 ml of 10^7 *S. mitis*
V					
	A	9	Palatal wound	Into palate	1 ml of 2×10^7 *S. mitis*
	B	41	Extraction of incisor	Into extraction wound	1 ml of 2×10^7 *S. mitis*

All rabbits were killed 3 to 7 days following the manipulative and inoculative procedures. The thorax was opened aseptically, the great vessels were clamped, and the heart was surgically opened and examined for vegetations on the endocardium and aortic valve. Vegetations were removed and cultured on blood agar plates. Smears were Gram-stained and the specimens inoculated into thioglycolate broth. Growth was subcultured into blood agar and mitis-salivarius agar and comparisons made with the test organisms colonially, microscopically, and morphologically. Samples of the palate at the injection site and granulations arising from the apexes of the extracted teeth were plated, harvested, homogenized, and serially diluted for quantification of the bacteria in infected extraction sites.

All of the catheterized animals had macroscopic cardiac vegetations ranging from 1 mm to almost total occlusion of the aortic valve. Infected vegetations were friable, aciniform, pale pink excrescences. Sterile vegetations were less friable, flatter, and paler. Catheterization, oral manipulation, and an initial inoculum of up to 10^6 organisms into freshly bleeding wounds were not enough to overwhelm the host's defense mechanisms and induce experimental endocarditis. Endocarditis was consistently obtained when the challenge was increased to 10^7 organisms per milliliter and the blood level of the organisms reached a calculated level of 10^5. Essentially complete infection of the experimentally induced cardiac vegetations was attained in Group V B, which most closely simulated the clinical situation of a human tooth with a periapical abscess.

Rizzo et al.[34] examined the pathogenicity of oral spirochetes in the hamster cheek pouch. Golden hamsters weighing approximately 100 g were anesthetized with Nembutal® and both cheek pouches exposed, washed with sterile saline, and stretched over a wide mounting board. Injections of *Borrelia buccalis* and of small treponemes were made into the pouch mucosa near the blind end. Approximately 3×10^5 bacterial cells suspended in 0.01 mℓ of fluid were introduced to form a bleb about 3 mm in diameter. Organisms were inoculated into one pouch and either sterile or spent medium into the contralateral pouch. Gross and microscopic examinations were made of specimens taken at intervals after inoculation.

Within 24 hr, there were well-defined abscesses in all of the pouches inoculated with organisms. The lesions were hemorrhagic, had a peripheral hyperemia, and contained a central core of thick yellowish-white exudate. After reaching maximum severity in 24 to 48 hr without ulceration, the lesions began to gradually regress. In 6 to 8 days, the purulent exudate had resorbed, leaving only hyperemia and induration; after several weeks, the only residues were small scars. Histologically, the well-delineated inflammatory exudate contained neutrophils, macrophages, spirochetes, and debris. Dark field examination of material from the 24-hr lesions disclosed actively motile spirochetes.

Hampp and Mergenhagen[35] produced spirochetal abscesses following i.v. inoculations of guinea pigs and rabbits with pure cultures of representative strains of the various morphologic types of oral spirochetes. Those used in this study were *B. vincenti, B. buccalis,* and small oral treponemes. One mℓ portion each of living cells, heat-killed cells, filtered culture supernatants, heat-inactivated filtered culture supernatants, and uninoculated sterile broth were inoculated intracutaneously into the shaved skin of animals. Guinea pigs weighing between 150 and 200 g were used in the early phase of the study. They were replaced later by New Zealand white rabbits weighing 1.5 to 2 kg because the rabbits proved to be more appropriate from the standpoint of severity of the lesions and the large amount of skin available for inoculation. The lesions were removed surgically at intervals and examined under dark field and light microscopy.

By injection of the pure strains of *B. vincenti* and *B. buccalis,* 9 intracutaneous abscesses were produced in guinea pigs and over 60 in rabbits. The minimum effective dose for initiating the lesions was 4×10^6 spirochetes. A dose of 6×10^7 cells was an effective and convenient inoculum for studying the infectious process that was produced. An injection of 1.0 mℓ containing 6×10^7 cells produced an immediate wheal 1.5 cm in diameter and 0.4 cm high, followed in 1 hr by a flattened ischemic area roughly proportional to the size of the wheal at the inoculation site. In 2 hr, the ischemic area was surrounded by hyperemia and edema. By 24 hr, there was gross evidence of abscess formation. The lesions usually persisted for a month without ulceration and drainage. Occasionally, there was spontaneous early ulceration and healing in 2 to 3 weeks. Others that ulcerated early became encrusted and remained as low-grade infections for up to a month. Every abscess contained a thick, purulent, off-white caseous material that did not exude spontaneously when the surface ulcerated.

Uninoculated finished medium, cell-free broth or heat-inactivated cell-free spent broth did not produce any lesions on inoculation. Spent broth containing spirochetes produced typical abscesses in every instance. An inflammatory response was elicited within 1 hr after infection that increased in amount to produce definite signs of early abscess formation by 12 hr. By 48 hr there was a beginning of fibroblastic activity around the abscesses while the spirochetes were concentrated in the upper and mid portions of the reticular layer of the dermis.

Swenson and Muhler[36] induced fusospirochetal infections in dogs whose tissue resistance was lowered by daily i.v. injections of Scillaren B, a glucoside with pharmacologic properties similar to those of digitalis. Injections of 0.08 to 0.1 mg/kg were used. The reaction to the first injection was a decrease in heart rate and increase in stroke volume. After the second injection, the dogs showed signs of lassitude, anorexia, sialorrhea, and regurgitation. These symptoms became progressively accentuated each day eventuating in extreme debilitation. Each dog had fusiform bacilli and spirochetes in the gingival sulcus before the injections began. The number of these organisms increased greatly after the third and fourth injection. Concurrently, a dark-brown-colored debris appeared on the posterior teeth. Gingival margins became acutely inflamed, had a putrid odor, and bled on slight provocation. By day 5 or 6, the mucous membranes of the mouth contained areas of hyperemia. These were usually located on the buccal mucosa adjacent to the posterior teeth and on the gingiva. Later the surfaces became granular and ulcerated. Smears from the ulcers showed spirochetes and fusiform bacilli.

Many of the oral Gram-negative bacteria and spirochetes contain lipopolysaccharide endotoxins. Rizzo and Mergenhagen[37] found that such endotoxins can produce mouth lesions when injected intramucosally in rabbits. The endotoxin used in these experiments was extracted from washed, acetone-dried cells of an oral strain of *Veillonella*. Single injections of 50 µg of a phenol-water preparation of the endotoxin into the palatal gingiva resulted in the development of gross abscesses lasting at least 10 days. Microscopic inflammation was evident for 14 days. Clinical signs during the first day were edema, erythema, and subepithelial white spots 1 to 3 mm in size. The spots consisted of collections of polymorphonuclear leucocytes infiltrating an area of necrosis corresponding to the exact location of the inoculating needle. Cellular infiltration peaked at 24 hr, followed by diffuse and focal necrosis in the lamina propria that formed abscesses. Early lesions contained areas of palatal bone resorption by osteoclastic activity. Regional lymph nodes showed neutrophilic infiltration and subsequent proliferative changes. In related experiments, the minimal preparatory dose for the local Schwartzman phenomenon in the mucosal test site was approximately 1.0 µg endotoxin.

Taichman and Courant[38] produced hemorrhagic necrotic lesions in the hamster cheek pouch and skin by i.p. administration of bacterial lipopolysaccharide (*E. coli* endotoxin) followed by local injection of catecholamines. Each animal received 10 µg *E. coli* endotoxin 1 to 3 hrs prior to anesthesia and injection of 1.0 µg epinephrine or 0.01 µg norepinephrine submucosally and intradermally. The intraoral injection was made into the blind end of the cheek pouch on one side. The opposite pouch was injected with a similar volume of saline and acted as control. A positive response was macroscopic evidence of tissue necrosis. Epinephrine produced positive responses in 6/6 animals; norepinephrine, in 4/6 animals. Similar lesions were induced by the administration of 10 µg *Leptotrichia buccalis* endotoxin systemically and of epinephrine or norepinephrine locally. The necrotic lesions showed engorgement of the vascular channels, and many of the vessels were disrupted. The vast majority did not contain thrombi. Hemorrhage was evident in all layers of the connective tissue.

Fungal Infections

Adams and Jones[39] detailed the life history of experimentally induced acute oral candidiasis in the rat. In this study, 42 Wistar rats were separated into seven groups. Each group contained six rats of the same sex and age. *Candida albicans,* Type A, was grown on Sabouraud's agar for 24 hr at 37°C. Surface growth was harvested in normal saline and washed once at 3500 rpm for 10 min. The organisms were resuspended in normal saline, and 10^8 cells in 0.25 mℓ were placed in the mouths of unanesthetized animals in Groups I through V. The other two groups served as controls. One rat from each group was killed at the end of each week for 6 weeks. Heads were fixed in 10% formalin, decalcified, blocked, paraffin-embedded, sectioned, and stained with hematoxylin and eosin and with PAS.

Five of the 30 rats inoculated with the test organism had histologic evidence of candidiasis. Infections nearly always occurred in the mandibular molar region under the alveolar crest. Typical lesions showed fungal hyphae penetrating the stratum corneum. During the first 2 weeks, the tissue response consisted of mild epithelial hyperplasia, followed in 1 to 3 weeks by parakeratosis and thickening of the stratum corneum. The outer cells were swollen by hydropic changes in cytoplasm and were separated from each other by small spaces. Acute inflammatory cells infiltrated the parakeratotic epithelium. The corium contained a minimal number of inflammatory cells. Invading organisms were always of the mycelial variety, with yeast cells seen only on the epithelial surface.

Subsequently, Jones and Russell[40] described the histopathology of chronic candidal infection in the rat's tongue in an experiment lasting 22 weeks. A total of 140 Sprague-Dawley male and female rats, average weight 200 g, was divided into two experimental groups of 60 each and a control group of 20 rats; all were housed in batches of 10. During the 1st week, animals in the experimental groups received a 0.1% aqueous solution of tetracycline HCl to drink at a level of 40 mℓ per rat per day. In the second week, they were inoculated orally with a 1.0 mℓ suspension of 6×10^7 *C. albicans* cells in the yeast phase on alternate days and given 0.01% tetracycline in the drinking water. One of the experimental groups drank the tetracycline-containing water throughout the experiment; the other for 1 week. Ten animals in each experimental group were sacrificed at the end of the 6th, 8th, 10th, 14th, 17th, and 22nd study weeks. Of the 20 controls, 10 drank tetracycline-containing water throughout the study period and 10 did not.

Mouth swabs yielded *C. albicans* from all of the animals in the experimental groups in the 1st week after inoculation. At the end of 22 weeks, the organism was recoverable from the mouths of 50% of the rats. *Candida* was not present in the mouths of any of the controls at any time. Infection was most frequent on the dorsum of the tongue, where the lingual papillae were replaced by a flat surfaced parakeratotic epithelium. Mycelial elements invaded the parakeratotic layer, but stopped short of the underlying stratum spinosum. Epithelial cells beneath the intact cornified layer were disorganized in arrangement and altered in appearance. Nuclei were large, and the nuclearcytoplasmic ratio was increased. Although mitoses and atypical cells with enlarged nuclei were common in the upper layers, the basal layer was always intact. Beneath the infected epithelium, the corium was excessively vascular and infiltrated with mononuclear cells. Many of the underlying muscle cells were shrunken and had basophilic cytoplasm. Some cells contained collections of large vesicular nuclei and prominent nucleoli occasionally arranged in rows or groups in the center of the cytoplasm. There was sarcolemmal proliferation and mononuclear cell infiltration between the muscle fibers and around small blood vessels. The changes produced a shallow, wedge-shaped area immediately below the infected epithelium, with the apex toward the center of the tongue.

The effects of prolonged candidal infections lasting 6 to 12 months on the rat tongue were elicited by Russell and Jones.[41] In this study, 60 Sprague-Dawley male and female rats received 0.1% tetracycline HCl in the drinking water for 1 week and 0.01% thereafter. During the 2nd week, 0.1 ml of a suspension of C. albicans in normal saline (6 × 10⁷ cells per milliliter) was placed in the rats' mouths on three occasions, after which a single dose was given weekly. Oral swabs were taken at the beginning of the experiment before inoculation with the fungus and at 2-week intervals thereafter for culture of the organism. After 6, 9, and 12 months, 20 rats were sacrificed. Long-term candidal infection in the rat tongue produced marked epithelial hyperplasia with keratin trapped between folds of epithelium. The epithelium showed atypia that were not carcinoma in situ. Degenerative changes were found in the superficial layer of the lingual musculature beneath areas of candidal infection. After a year, the damaged muscle was replaced by hyalinized fibrous tissue.

Jones and Russell[42] studied experimentally produced candidiasis in weanling rats. Four litters of 12-day-old rats housed with their mothers were tested. Five of ten rats in Litter A and six in Litter B were inoculated orally with 0.1 ml saline containing 10⁸ C. albicans cells in the yeast phase. Mouths were swabbed on the 5th day and the swabs placed in Sabouraud's broth containing antibiotics. Cultures were incubated for 10 days at 37°C and examined for yeastlike cells. After swabbing, two inoculated and two uninoculated rats from each litter were killed, the heads removed, decalcified in toto, blocked, processed, sectioned, and stained with either hematoxylin and eosin or PAS. The remaining rats were swabbed and killed on the 12th day.

Each of the eight rats in Litter C and the seven in Litter D were inoculated orally with 0.01 ml saline containing C. albicans in the mycelial phase on the 1st, 3rd, and 5th day. On the 8th day, all rats were swabbed and Litter C killed. Litter D was reinoculated on the 8th, 10th, and 12th day, swabbed and killed on the 15th day. All cultures were incubated for 7 days and subcultured on Candida medium (oxoid) if yeast cells were seen in the broth. None of the rats in Litters A and B monoinoculated with the yeast form harbored the organism at swabbing or had histologic evidence of oral Candida infection. In Litter C, six of the eight rats harbored the organism and one had histologic evidence of infection. Infancy per se is thus not an important predisposing factor in experimental oral candidiasis in the rat. The infant rat is not receptive to Candida in the yeast form, but is receptive to the mycelial phase on repeated exposure.

Buditz-Jorgensen[43] examined the immune response to C. albicans in monkeys with experimental palatal candidiasis. In this study, 14 adult Macaca irus monkeys, 8 female and 6 male, were swabbed orally for cultural demonstration of C. albicans. Both upper first molars were extracted from each animal and alginate impressions made of the maxilla. Heat-cured acrylic plates were fabricated from plaster casts. Plates were retained in the mouth mechanically by extensions into the edentulous spaces. They were worn throughout the study following healing of the extraction sites. An experimental infection was produced by inoculating 100 mg wet weight of C. albicans, Type A, under the plates. Infection was confirmed by microscopic examination of PAS-stained palatal smears counterstained with hematoxylin. Half of the animals were subjected to immunosuppressive therapy (azathioprine) for 2 weeks before and 2 weeks after the infection. The course and development of cellular and humoral immune responses to C. albicans were assessed, using leucocyte migration tests and agglutination reactions, respectively.

In the normal animals, leucocyte migration inhibition was significant from the 1st week to 5 months after infection. Cellular hypersensitivity developed concomitantly with clearing of the infection, while antibody was not yet detectable. As the antibody titer rose, cellular hypersensitivity declined. In the azathioprine-treated animals, the infections persisted and cellular hypersensitivity did not develop until 1 to 3 weeks after drug treatment was discontinued. Humoral antibody response was early and strong. Inflammation in the azathioprine-treated monkeys was reduced compared to normal animals. Cellular hypersensitivity is apparently of paramount importance in the development of host resistance to experimentally induced superficial candidiasis.

Viral Infections

Rizzo and Ashe[44] followed the development of experimental herpetic ulcers in the rabbit oral mucosa clinically, culturally, and histologically. The strain of *Herpes simplex* used in the study was isolated from the lower lip of a subject with recurrent herpes labialis. Multiple palatal mucosa sites were inoculated with 10^{5-7} plaque-forming units of virus in New Zealand white rabbits weighing 1.5 to 3.0 kg. Injection points were erythematous and slightly edematous during the 1st day. Small white elevations or plaques appeared in the next 24 hr. Removal of the plaques by vigorous cotton swabbing left irregular bleeding ulcers. Frank ulcers one to several millimeters in diameter with a grayish-white nondepressed base and marked peripheral erythema were present by the end of the 2nd day. By days 3 and 4, the inoculation sites contained clusters of ulcers with a central lesion surrounded by one or more satellite lesions. The base of the ulcer was slightly depressed and the epithelial rim sharply outlined, producing a punched-out appearance. Between the 4th and 6th day, the lesions coalesced to form single ulcers 4 to 6 mm in diameter that began to contract between days 6 and 9 and reepithelialized between days 9 and 12.

Histologically, the inoculation sites showed signs of herpes infection throughout the 6-day period of ulcer development. Connective tissue fibroblasts contained distinct nuclear inclusions as early as 8 hr after inoculation. Definite signs of herpetic infection became manifest in the mucosal epithelium during the latter part of the 1st day and remained evident for 6 days. These included ballooning degeneration of the cytoplasm, eosinophilic nuclear inclusions, and margination of the chromatin to the nuclear membrane. The absence of the virus after day 6 coincided with the end of the clinical ulcer enlargement and signified the end of active infection.

Cytomegaloviruses belong to the herpes virus family and cause chronic infections in man and animals. Cytomegalovirus has caused salivary gland infections in children, adults, and mice. The characteristics of the ultrastructural lesions in the submaxillary salivary glands of mice during chronic murine cytomegalovirus infection were delineated by Henson and Strano.[45] Male C_3H/He and DBA/2 mice, 6 to 8 weeks of age, were inoculated i.p. with 0.25 mℓ virus suspension prepared by homogenizing salivary glands of 3-week postinfected C_3H mice. Virus suspension was prepared initially as a 10% homogenate in saline and diluted 10^{-2} before injection. Following infection, mice were sacrificed twice a week for 9 weeks and the submaxillary glands removed and prepared for examination by light and electron microscopy. Some mice were also given 0.5 mg cortisone i.p. three times a week beginning 10 days after the infection.

During the infection, intranuclear inclusion bodies appeared predominantly in seromucous-secreting cells, occasionally in mucous-secreting cells, and rarely in serous-secreting cells. Virus was synthesized in the nucleus of the acinar cells. After passage into the cytoplasm, virus was located in large vesicles derived from the Golgi apparatus. The vesicles which were PAS-positive migrated to the apex of the cell and released virus into the acinar lumen or canaliculi. Eventually, lymphocytes infiltrated the interstitium and surrounded the basal lamina of the acini-containing infected cells.

In acini encompassed by lymphocytes, infected cells and morphologically normal cells degenerated simultaneously, producing a small focus of necrosis. Physical contact between lymphocytes and necrotic cells did not occur, as an intact basal lamina was always interposed between them. Degeneration of infected cells coincided with a decrease in virus titer of the salivary glands. Degeneration of infected and normal acinar cells also occurred in DBA/2 mice that lack the fifth component of complement. In mice conditioned with cortisone to suppress inflammation, neither infected nor normal cells degenerated. Lymphocytes responding to the infection apparently released a cytotoxic substance that diffused into the acini and caused indiscriminate necrosis of both normal and infected cells.

Joseph et al.[46] demonstrated that murine cytomegalovirus induces not only intranuclear inclusions in acinar epithelial cells of mouse submandibular gland, but also two distinct morphologic types of viral intracytoplasmic inclusions. Young female CFW strain mice weighing between 15 and 50 g were inoculated s.c. with 0.2 ml of a pooled homogenate containing infectious murine cytomegalovirus. Submandibular glands were harvested from mice sacrificed at intervals ranging from 1 hr to 324 days. Two additional groups of mice served as controls. One control group of three mice received 0.2 ml of pooled murine cytomegalovirus-infected submandibular gland homogenate heated at 78°C for 30 min. to render the virus noninfective by s.c. injection. The other control group contained three normal noninfected young CFW female mice. All submandibular gland specimens were prepared for examination by electron microscopy.

In the newly infected cells, the first stage of virus replication was the formation of a single shell-like structure within the nucleus. Shortly thereafter, a second shell developed within the first. The inner shell contained large amounts of DNA and appeared to condense, forming a densely staining viral core that was present in most of the cytoplasmic viral particles. Virus particles were also located near the Golgi apparatus. Light types of cytoplasmic inclusions preceded the formation of the dark type. The dark type appeared to be a later stage or mature form of murine cytomegalovirus intracytoplasmic inclusion body and contained a greater number of virions than the light type. The findings point to the possibility of viral replication or maturation occurring within the cytoplasmic inclusion during the later stages of the infection.

Many of the childhood viral exanthematous diseases leave telltale signs in the teeth in the form of hypoplastic enamel and dentin. Kreshover and Hancock[47] elicited the pathogenesis of abnormal enamel formation in rabbits inoculated with vaccinia virus. Six rabbits were injected i.v. with dermal lymph strain of vaccinia in dosages previously demonstrated to produce a sublethal generalized infection. Five rabbits with a past history of respiratory infection were included for comparative purposes. Eight healthy uninfected animals served as controls. The control animals were sacrificed between 30 and 52 days of observation, the vaccinia inoculated group between 34 and 90 days, and the respiratory group between 32 and 92 days. Radiographs were made of the jaws and both the maxillas and mandibles were prepared for histologic examination.

Rabbits infected with vaccinia virus showed marked developmental dental defects when compared with the other groups. Incisal eruption rate was significantly slowed, accompanied by development of hypoplastic enamel and dentin. Normal serum calcium and phosphorus levels in the vaccinia-infected rabbits indicated that the dental changes were caused by the virus rather than by hormonal or dietary deficiencies.

Kreshover et al.[48] determined the effects of vaccinia infection in pregnant rabbits on maternal and fetal dental tissues. Vaccinia virus in 1 ml doses of titers ranging from 10^{-4} to 10^{-6} were injected into the ear veins of 46 female California and New Zealand albino rabbits ranging in age from 6 to 11 months. Of the 33 virgins, 14 were injected 6 to 16 days prior to mating; the remainder, after mating. Of the group, 12 were inoculated during the first trimester of pregnancy (1 to 10 days), 4 during the second trimester (11 to 20 days), and 3 during the third trimester (21 to 30 days). Two rabbits in each group were permitted to complete pregnancy without disturbance. Caesarean sections were performed in the others at different stages of gestation. Young, spontaneously delivered at term were killed at intervals between 1 and 60 days of age. Their mandibles and maxillas, along with those of the fetuses removed by section, were prepared for histologic study. Blood and tissues were tested for the presence of vaccinia virus.

Striking gross and microscopic dental defects were found in the incisor, premolar, and molar teeth of the pregnant rabbits infected with vaccinia. Hypoplastic lesions in the incisors became grossly evident approximately 160 days after inoculation. Linear grooving and pitting of the coronal surfaces were characteristic. Histologically, both the enamel and dentin showed intermittent abnormal matrix formation and hypocalcification. Similar changes were noted in the premolars and molars. No dental abnormalities were found in the fetuses or delivered young of mothers infected with vaccinia, regardless of stage of gestation when the mother was inoculated.

Kreshover and Hancock[49] explored the effects of experimentally induced lymphocytic choriomeningitis on the course of pregnancy and on murine dental tissues. Lymphocytic choriomeningitis virus was injected i.p. into 153 pregnant Swiss mice; 20 others were similarly inoculated with a heat-killed suspension of virus, and 30 served as breeder controls. The viral inoculum was prepared by injecting 6- to 8-week-old normal mice intracerebrally with a suspension of blood from five congenitally infected mice. Following appearance of the characteristic signs of the disease in 6 to 8 days, the animals were sacrificed and their brains aseptically stored in 60% neutral glycerol at 4°C. Macerated brain tissue was suspended in saline and diluted to 10^{-3} as needed. Doses of 0.25 mℓ of this dilution were administered to the 153 bred mice; 68 were inoculated during the first trimester (1 to 6 days), 63 during the second trimester (7 to 13 days), and 22 during the last week of pregnancy (14 to 21 days). The young of the animals that delivered spontaneously at term were sacrificed at 1 to 28 days of age. Maxillas and mandibles were removed for histologic study, and blood samples were inoculated into normal mice to test for the lymphocytic choriomeningitis virus.

The course of pregnancy was adversely affected by the infection. There was a high percentage of abortions, resorptions, and other instances of nonprogression of pregnancy when the mother was inoculated during the early stages of gestation. The duration of the viremia following virus inoculation in the adult mice was brief. Pulpal hyperemia and hemorrhage were consistent findings in the incisor and molar teeth of the inoculated mice. Odontoblastic degeneration was noted occasionally. Dental abnormalities were relatively infrequent in the young of infected mothers. When present, they were typified by disturbed amelogenesis rather than by the pulp changes found in the mothers.

EXPERIMENTAL OSTEOMYELITIS OF MAXILLA AND MANDIBLE

Lewin-Epstein and Neder[50] devised a method for the production of chronic osteomyelitis in the maxilla of the rabbit. Nine rabbits weighing 2 to 3 kg were anesthetized and a 3 mm hole was drilled down to the cancellous bone through a raised mucoperiosteal flap in the diastema between the maxillary incisor and molars. A sterile cotton pellet was inserted into the hole and 0.02 mℓ of fresh inoculum (24-hr culture of hemolytic coagulase positive *Staphylococcus aureus*) was slowly injected into the bone beyond the cotton that prevented spillage by absorption. The hole was sealed with sterile beeswax and the wound closed with silk sutures. The implantation of 100,000 cells was made in six animals, and of 50,000 cells in three animals. X-rays of the involved bones were taken pre- and postoperatively. At sacrifice, the bone was removed and processed for histologic examination. Eight of the nine animals developed chronic osteomyelitis of the maxilla by this procedure.

Triplett et al.[51] developed an experimental model for the study of osteomyelitis in the mandible of rabbits. New Zealand white rabbits weighing between 4 and 5 kg were anesthetized and the right mandible exposed by an extraoral surgical approach. Osseous cuts were made to separate the medial and lateral cortical plates. The defect exposed the root surfaces and created a compound sagittal mandibular fracture that communicated with the oral cavity along the roots of the teeth. Medullary tissue was thoroughly cauterized to diminish local blood supply and to provide a coagulum for bacterial growth. Eight rabbits received an inoculum of 0.25 m*l* of a 24-hr brain-heart infusion broth culture of *S. aureus*; ten rabbits were inoculated with 0.25 m*l* of a 24-hr broth culture of *Bacteroides melaninogenicus* grown in peptone yeast glucose broth supplemented with vitamin K and heme. Wounds were closed without drainage and the animals isolated and maintained on normal diets. Five rabbits surgically treated but not inoculated served as controls.

Criteria for confirmation of osteomyelitis were a chronic clinical infection demonstrable by visual inspection at open biopsy and positive bacterial cultures 3 months after inoculation. Two of the rabbits inoculated with *S. aureus* met the criteria, as did all ten inoculated with *B. melaninogenicus*. The predominant oral isolates recovered at biopsy were mixed cultures of aerobic and anaerobic organisms. The *B. melaninogenicus* model demonstrated many of the clinical aspects of the human manidublar osteomyelitis, notably, swelling, drainage, and mixed bacterial flora in the exudate. All controls healed normally without infection.

EXPERIMENTAL SIALADENITIS

Acute sialadenitis can be induced by instillation of an antigenic solution into the parotid gland of immunized rats. Sialadenitis also develops after intraglandular introduction of an antiserum to rat blood or to basement membranes. The inflammatory response in these models has been attributed to the formation of immune complexes in the glands that activate the complement system initiating release of mediating substances.

Sela et al.[52] described the response of the parotid gland of sensitized and nonsensitized rats to multiple instillations of antigen administered at weekly intervals. The parotid gland and knee joint of immunized albino rats of both sexes weighing between 120 and 200 g were challenged with sensitizing antigen. Sialadenitis developed in over half the rats after one to three challenges through the parotid gland duct. Most of the glands were normal after five, seven, and eight intraductal challenges despite the presence of precipitating antibodies in the serum and hemagglutinating antibodies in parotid saliva. A few animals developed chronic destructive adenitis. In contrast, the incidence and severity of synovitis increased with the number of intra-articular injections of antigen. The nonreactivity of the gland was antigen specific. Acute adenitis resulted from intraductal instillation of a second antigen following multiple challenges with the first of two antigens to which the rat was sensitized.

White and Casarett[53] induced experimental autoallergic sialadinitis in the submandibular gland of rats by immunization with allogeneic submandibular gland homogenate and adjuvants. The recipients were male Lewis rats 9 to 12 weeks of age; the donors were sexually mature and sexually immature female Wistar rats. Donor submandibular glands were excised, cleaned of fascia, and homogenized with an equal volume of phosphate-buffered saline. The homogenate was emulsified with an equal volume of complete Freund's adjuvant containing 0.4% *Mycobacterium butyricum*.

Rats were immunized under light ether anesthesia and injected with 0.01 m*l* of the submandibular gland adjuvant emulsion into the plantar surface of each foot. Additionally, 0.10 m*l* of *Bordetella pertussis* vaccine (2×10^{11} cells/m*l*) was injected s.c. in the dorsal surface of each foot. Recipient animals were sacrified 8 to 56 days after immunization. Organs were prepared for histologic analysis and scored for degree of inflammatory involvement. Control groups were comprised of animals that were either nontreated or given adjuvant only.

Immunization of the Lewis rats with a saline homogenate of mature or immature allogeneic submandibular glands emulsified in CFA and containing *B. pertussis* vaccine provoked an autoallergic sialadenitis in the submandibular gland. The lesions were first detected 8 days after immunization and appeared as small foci of infiltrating mononuclear cells adjacent to ducts. By 12 to 14 days, infiltration with mononuclear cells was widespread and parenchymal distribution was extensive. Only some small tubules were spared. Mitotic activity appeared to be limited to the inflammatory cells. By 21 days, there was a regeneration of the acinar elements and proliferation of tubules with mitotic activity. At 28 to 56 days, the glands were histologically normal except for occasional foci of inflammatory cells.

White[54] created a similar experimental autoallergic lesion in the rat parotid gland. Young male Lewis rats were immunized with a mixture of parotid gland homogenate and adjuvants. Parotid gland tissue from adult female Wistar rats was homogenized with phosphate-buffered isotonic saline and emulsified with an equal volume of CFA. The emulsion was injected intradermally into each foot of the rat. The rats were also given s.c. injections of *B. pertussis* in the dorsal surface of each foot. Control animals were either not treated or received adjuvants only. The animals were killed 14 days after immunization and the parotids removed and examined histologically. Nine of ten animals developed autoallergic parotitis. Lesions were manifested by parenchymal degeneration and mononuclear cell infiltration, primarily with lymphocytes and macrophages. The large incidence, high grade of the lesions, complete specificity of the target organ, and nature of the infiltrating cells were consistent with a cell-mediated immune process.

Boss et al.[55] developed a model of immunologically mediated chronic destructive sialadenitis precipitated by daily antigenic challenge. The right parotid duct of albino rats of either sex was instilled five to eight times with bovine serum albumin on a daily basis. Acute parotitis developed in 8 of 20 nonsensitized animals. Chronic destructive sialadenitis with parenchymal destruction and mononuclear cell infiltration was generated in the preimmunized rats. Morphologic changes resembled those found in the salivary glands of patients with Sjogren's syndrome. The earliest manifestation was a scattering of lymphocytes about the ducts. Subsequently, the accumulations of lymphocytes enlarged and extended into the lobules. Hyperchromasia of acinar cell nuclei and acinal involution followed. Ultimately, the entire parenchyma, except for the ducts, was replaced by inflammatory infiltrates.

Whaley and MacSween[56] produced immune sialadenitis in guinea pigs using three different adjuvants. The test series was composed of female Duncan-Hartley guinea pigs, 4 months of age, weighing between 450 and 500 g. Homologous guinea pig salivary gland was homogenized, purified, and injected into the hind footpad and s.c. interscapular region of each animal. The amount injected was 0.5 m*l* of supernatant from a centrifuged mixture of homogenized gland and physiological saline made up to 1.0 m*l* with either CFA or phosphate-buffered saline (pH 7.2) before injection. Carbonyl iron (250 mg per animal) was injected with antigen and phosphate-buffered saline solution as the second adjuvant. *B. pertussis* vaccine (0.2 m*l* per animal) was instilled into the right forefoot pad at the time of immunization as the third adjuvant. Control animals were given 1.0 m*l* of the antigen-phosphate buffered suspension.

Periductal inflammatory infiltrates were commonly produced when two adjuvants were given together but were unusual when each adjuvant was used alone. Ensuing lesions contained a plasma cell and lymphocyte infiltrate focally distributed near the duct. Some animals had extensive and diffuse lesions in which the chronic inflammatory cell infiltrate involved acinar cells and acinar loss. One of four animals given either carbonyl iron or *B. pertussis* vaccine developed sialadenitis. Lesions were found in all animals given both carbonyl iron and *B. pertussis* vaccine. In only one of five animals given the three adjuvants was there a mild degree of sialadenitis. The most severe lesions occurred when carbonyl iron and *B. pertussis* vaccine were used together.

EXPERIMENTAL ORAL ULCERS

Although the role of psychologic factors in the etiology of peptic ulcers has been investigated repeatedly, rarely have such influences been implicated in diseases of the oral mucosa. Whittaker and Wilson[57] studied the effects of physical restraint on the production of oral and gastric ulcers in three different strains of rats ranging in age from 6 to 12 weeks. Used in the experiments were 64 male and 64 female rats each of the Sprague-Dawley, Wistar, and Wistar hooded strains. The 128 rats of each strain were divided into four groups of 16 males and 16 females with eight experimental and eight control rats per group. The four groups from each strain were tested at mean ages of 6, 8, 10, and 12 weeks. Restraint was obtained by application of plaster of Paris bandages to the rear half of the body, abdomen, and hind legs while the animal was under ether anesthesia. The casts restricted general body movement without impairing respiration. After 24 hr of restraint, the rats were killed and the stomachs removed, washed with warm water, and inflated with formol saline until fully distended. Severity of gastric ulcers was evaluated macroscopically and microscopically. To analyze for oral ulcers, the rats were decapitated and the heads fixed in formol saline. The jaws were propped open, the tongue and lower lip dissected out, and the lower jaw removed to expose the hard and soft palates, which were examined under a dissecting microscope for breaks in the continuity of the oral epithelium. The search was aided by use of 2% fluorescein solution. Some of the ulcers were excised and the tissues processed for histologic examination.

There were no significant sex differences in ulcer incidence. Increasing age was associated with a diminution in frequency of gastric ulceration. The number of gastric ulcers was highest in the Wistar strain, less in the Wistar hooded, and least in the Sprague-Dawley animals. Oral ulcers in numbers significantly greater than in the controls were found only in Sprague-Dawley rats. Oral ulcer incidence was not affected by age.

Zackin and Goldhaber[58] produced oral ulcers in rats by periodic reduction of the lower incisors. In this study, 64 female albino rats weighing approximately 125 g were divided into four groups. Group I was untreated; Group II was injected with cortisone; Group III had the lower incisors cut down periodically to the gingival margin; Group IV was given cortisone injections along with the incisor cutdowns. Animals in Groups II and IV received 2 mg cortisone daily s.c.; animals in Groups III and IV had the lower incisors shortened twice weekly. All were fed ground stock diet *ad libitum*. Each was examined after 1 month for oral lesions, killed, and posted for histologic study.

Repeated shortening of the lower incisors produced deep ulceration in the labial mucosa, gingiva, and tongue, regardless of cortisone administration. No lesions were found in animals with intact lower incisors. The ulcers were chronic and nonspecific based by granulation tissue containing lymphocytes, monocytes, and plasma cells and surfaced by polymorphonuclear cells, fibrin, and necrotic debris. Cortisone did not alter either the gross or microscopic appearance of the lesions. Neither injections of penicillin-streptomycin solution nor extension of the experimental period to 2 months affected either incidence or extent of lesions.

Taylor et al.[59] created gingival ulcerations in beagle dogs by the administration of radionuclides. Young adult, pure-bred beagles in good health with complete permanent dentition were given single i.v. injections of Ra^{226}, Ra^{228}, Th^{228}, Pu^{239}, or Sr^{90}. Others received seven injections of Ra^{226} at 60-day intervals.

Radiation-induced oral conditions, resembling those described in watch dial painters, were produced in several dogs. Ulcerative gingival lesions developed in those receiving relatively high doses of Ra^{228} and of Th^{228} by single injection, and of Ra^{226} by multiple injections. Lesser symptoms without ulceration were seen in dogs given single injections of Ra^{226} and Pu^{239}. Ulcerations did not develop in those given Sr^{90}. The gingival lesions differed from those caused by natural factors, with the earliest change being a focal necrosis preceded by a minimal inflammatory reaction. Gingival ulcers often appeared in the absence of any dental or periodontal disease. Reparative tendencies were absent or minimal. Pathogenesis of the gingival ulceration appeared to be secondary to underlying bone changes. By the time the epithelial layer had perforated, bone could be seen at the base of the ulcer. Such bone was devoid of periosteum, abnormally white, relatively insensitive, and microscopically necrotic. The major features of the experimentally produced disease were strikingly similar to radiation necrosis of the mandible following irradiation for oral malignancies.

ORAL RESPONSES TO EXPERIMENTAL BLOOD DYSCRASIAS

Attstrom and Egelberg[60] studied the effects of experimental leucopenia on gingival inflammation in dogs. Nine beagles (7 male, 2 female) aged 12 to 13 months, weighing 11 to 15 kg, were fed a soft gingivitis-causing diet. Leucopenia was created by i.v. injection of nitrogen mustard at doses of 0.8 mg/kg of body weight on day 0 and 0.4 mg/kg body weight on day 7. Venous blood samples were taken daily for white cell counts and differentials. Samples of leucocytes from 24 buccal gingival crevices were obtained by gently inserting a styroflex film strip into each crevice. Total area of the styroflex film showing leucocytes was calculated and differential counts performed. Leucocyte areas on all 24 films were added for each dog. Differential counts were made on films from one gingival crevice in each quadrant from five dogs. Gingival fluid was measured on unstained strips by determining the length of the moistened part to the nearest millimeter. Values for all 24 strips were totaled for each dog. Acid phosphatase activity in the crevice sample was assessed in each quadrant. Samples were obtained from day -7 to day $+24$.

Maximal leucopenia occurred at days 7 to 10. During leucopenia there was a reduction in crevicular leucocytes, gingival fluid, and acid phosphatase activity. The values returned to normal in the postleucopenia period indicating that (1) crevicular leucocytes may contribute to the enzyme milieu in the gingival crevice, and (2) lysosomal enzymes from crevicular leucocytes may influence the tendency of dentogingival vessels to undergo increased permeability during chronic gingival inflammation.

In a companion study, Attstrom et al.[61] instigated leucopenia in six beagle dogs with chronic gingivitis by injections of heterologous antineutrophilic serum. During leucopenia, the number of leucocytes in the gingival crevice and gingival tissues was reduced, as was the amount of gingival fluid. Acid phosphatase, hyaluronidase, and protease activity in the crevicular samples decreased in parallel with the reduction in neutrophils. All parameters returned to normal during the postleucopenic period. These studies support the proposition that enzyme release from neutrophils at the dentogingival junction may induce vascular damage in periodontitis.

Later Attstrom and Schroeder[62] elicited the influence of experimental neutropenia on initial gingivitis in dogs. Experimental neutropenia was induced in three healthy litter mate beagles 1 year of age by repeated injections of rabbit antineutrophil serum. The antineutrophil serum was raised in rabbits by injecting neutrophils collected from glycogen-precipitated peritoneal exudates in beagles. The serum was administered s.c. in the back of the neck in doses of 0.5 to 2.0 mg/kg, depending on response to the previous level. Antineutrophil serum was first given approximately 2 weeks after gingival normality was attained by scaling, polishing, and tooth cleaning. During neutropenia, microbial plaque was permitted to form on the teeth. Injections were repeated on days 1, 2, and 3, and samples of crevicular leucocytes and gingival fluid were collected on days 0 and 4. Block biopsies of buccal gingiva taken on day 4 from selected maxillary and mandibular sites were processed for histologic examination. Selected semi- and ultrathin sections were used for histometric and stereologic analyses by light and electron microscopy.

Gingival fluid increased from day 0 to day 4 while junctional leucocytes increased in one dog only. Very few neutrophils were found in the junctional epithelium. Granulocytes and macrophages were diffusely scattered in the coronal connective tissue adjacent to the junctional epithelium. The gingival collagen appeared displaced by the inflammatory cells rather than dissolved. Neutrophils may contribute to the loss of collagen during initial gingivitis in dogs and seem to be important in limiting subgingival plaque growth along the tooth surface.

Recently, Andersen et al.[63] determined the effects of experimental neutropenia on oral wound healing in guinea pigs. For the wound healing studies, 16 male guinea pigs weighing 400 to 500 g were used; 12 animals for a source of neutrophils as antigen in the preparation of antineutrophil serum. Antineutrophil serum was administered i.p. at a dose of 3 to 8 mg/kg/day. Injections were started on the day before wounding and continued daily. The animals were anesthetized with Nembutal® 24 hr after the first injection, and standard excisional wounds were prepared in the palatal mucosa. Test animals were sacrificed at 6 hr and 1, 2, and 5 days after wounding. Tissue blocks containing the wound area were fixed in Karnowsky fluid adjusted to pH 7.4 with phosphate buffer, postfixed in 2% tetroxide, dehydrated in alcohol, embedded in Epon®, and sectioned serially at 1 μ. Corresponding excisional wounds in 15 normal animals were used as controls. Sections were evaluated for the progress of epithelial wound repair and for occurrence and location of inflammatory cells and bacteria by electron microscopy.

In the animals treated with antineutrophil serum, neutrophilic granulocytes disappeared from the wound cavity while the other inflammatory cells remained unchanged. These animals had bacterial invasion deep into the wound cavity. Epithelial repair occurred at a deeper level in the wounds of the neutropenic animals compared to normal animals. Rate of reepithelialization was the same in the neutropenic and control wounds. Neutrophils in oral wounds constitute the principal host defense against microorganisms and probably, indirectly, determine the direction and level of epithelial cell migration.

Primary polycythemia or polycythemia rubra vera is frequently accompanied by prominent oral manifestations. These include a red or purple discoloration of the tongue and oral mucosa, purple to red swollen gingiva, distention of the ranine veins in the sublingual space, severe petechiae in the oral mucosa, and bleeding tendencies. Since there have been no reports of the histopathologic changes in the oral tissues in this condition, Lopez and Miller[64] studied the mouth structures in rats subjected to cobalt-induced polycythemia. In that study, 50 Wistar rats weighing between 70 and 80 g were divided into a control group of 18 and an experimental group of 32. The latter were made polycythemic by the addition of 0.1% cobalt nitrate to the drinking water. Erythrocyte counts were made on tail blood at the beginning of the experiment and after one and four study months. Hematocrits were determined from blood obtained at necropsy. Histologic studies were performed on the viscera and oral structures.

There was a definite increase in red cell count and in hematocrit values in the group consuming the cobalt-containing drinking water. The red cell count reached a mean of 11.51 M/mm^3 after 1 month and 12.9 M/mm^3 after 4 months in the polycythemic animals. Mean hematocrit was 63 in the test and 47 in the controls at termination of the study. At necropsy, the polycythemic animals had reddening of the skin, particularly in the nose and ears, hepatomegaly, and splenomegaly. Histologically, there was moderate dilatation of the blood vessels and mild congestion of the spleen and liver. The only polycythemia-associated histologic change in the oral and paraoral structures of test animals was vascular dilatation.

EXPERIMENTAL ORAL LEUKOPLAKIA

Lingual leukoplakia has been produced experimentally in hamsters by Marefat and Shklar.[65] In this study, 45 male and female young adult Syrian hamsters were assigned to three test groups of 15 animals each; three groups of 10 hamsters served as controls. In Test Group I, the right lateral border of the middle third of the tongue was scratched three times a week with a Number 9 root canal broach while the hamster was under light ether anesthesia. This was followed by application of a 0.5% solution of 9,10-dimethyl-1,2-benzanthracene (DMBA) in heavy mineral oil with a Number 4 sable brush. Test Group II was subjected to the same protocol except that the DMBA was dissolved in acetone. In Test Group III, DMBA in acetone was applied three times a week without preliminary trauma. The first control group was subjected to trauma alone, the second to trauma followed by acetone application, and the third was untreated.

Test Group I developed erosive ulcerations in 1 to 2 weeks, and hyperkeratosis in 12 to 13 weeks. Lesions in Test Group II progressed from erosive ulcerations after 1 to 2 weeks to rough irregular white patches resembling clinical leukoplakia in 4 to 6 weeks. Test Group III evidenced leukoplakia in 11 to 12 weeks. Leukoplakia was also found in all animals in the first control group at 12 to 16 weeks, indicating thereby that the trauma alone was sufficient to produce oral leukoplakia in the hamster.

Craig and Franklin[66] developed a model for the production of epithelial hyperplasia and hyperkeratosis in hamster cheek pouch mucosa. Cheek pouches of 4- to 6-week-old male Syrian hamsters were painted with 50% turpentine in liquid paraffin (v/v) three times a week for periods ranging from 1 day to 16 weeks. The lateral wall of each pouch was retracted and the turpentine solution applied by six strokes of a Number 4 sable hair brush to the medial aspect of the pouch just posterior to the anterior vein. Cheek pouches from five nonpainted hamsters matched for age and sex served as controls. The animals were killed with Nembutal®. Tissues from the site of painting were removed, fixed in formol-acetic acid-alcohol and processed for paraffin-embedding. Sections were stained with hematoxylin and eosin. The pouch mucosa was examined histologically and measurements of epithelial thickness were made with a calibrated eyepiece micrometer.

An increase in epithelial thickness was first observed 48 hr after a single painting accompanied by inflammatory changes in the epithelial and connective tissues. Maximal epithelial thickening occurred after 9 weeks of thrice-weekly painting. Cheek pouches of animals without further treatment for up to 1 year after the 9 weeks' painting were indistinguishable from the controls. During the 9 to 16 week period of painting, keratin thickness remained relatively unchanged, but the epithelium decreased in thickness and stabilized at about twice the control value. The epithelial response to turpentine was that of a reversible benign epithelial hyperplasia with hyperkeratosis. The model had no premalignant connotation. Histologic changes in the turpentine-treated hamster cheek pouches were similar to those produced by chemical irritants and by friction.

Capodanno and Hayward[67] compared the therapeutic efficacy of podophyllum resin and salicylic acid in eradicating hyperkeratosis from oral mucous membranes. DMBA was used to induce focal hyperkeratosis in hamster buccal mucosa. A 0.5% solution in mineral oil was placed with a camel's hair brush on the mucosal surface of the cheek pouch of 15 unanesthetized 3-week-old Syrian albino hamsters. An estimated 31.3 mg of drug was applied to the pouch once a week for 12 weeks. The hamsters were allocated to four groups. Two had cheek pouches painted bilaterally with DMB; two had unilateral treatment.

Clinical evidence of hyperkeratosis was apparent by the 12th week. Eight weeks after the last application of DMBA, five animals with bilateral DMBA were given 20% salicylic acid unilaterally; six, 20% podophyllum unilaterally. In four hamsters with unilateral DMBA treatment, 20% concentrations of each keratolytic agent was applied bilaterally to two animals. The keratolytics were applied to anesthetized animals on a schedule of 1 week between treatments one and two, 10 days between treatments two and three, and each week thereafter for 4 weeks. After initial salicylic acid induced desquamation, ulceration, and bleeding, there was a significant decrease in amount of hyperkeratosis. Alterations in keratinization were manifested by a decrease in opacity of the keratin layer and by a disrupted stratum granulosum. In contrast, podophyllum had no significant keratolytic effect on the experimentally induced DMBA lesions.

EXPERIMENTAL ODONTOGENIC CYSTS AND GRANULOMAS

Odontogenic keratocysts have an aggressive growth potential, a tendency towards multiplicity, and a marked propensity to recur or to persist, properties not possessed by any other form of odontogenic cyst. The pathogenesis of such cysts is uncertain. Suggested origins are offshoots of the dental lamina before odontogenesis or remnants of the dental lamina after odontogenesis. In the course of transplantation studies in mice by Bartlett et al.[68] using molar tooth germs, cysts containing keratin were produced during development of the transplanted teeth. First maxillary molars from 2-day-old C57Bl/10 mice neonates were carefully removed and transplanted to a subcapsular site in the kidney of anesthetized adult isologous mice. Nephrectomies were performed on the recipient mice after 1 to 180 days posttransplantation. The kidneys were fixed in 10% neutral formol saline, decalcified, and processed for histologic examination. Sections were stained with hematoxylin and eosin, and some were subjected to special technics for demonstrating keratin or keratinlike material. As the tooth germs were forming teeth, keratin-producing cysts also developed in the epithelium of the enamel organs. By 120 days posttransplantation, many of the cysts had completely enveloped the crowns of the developing teeth. These studies show that tooth germ epithelium was capable both of cystic change and keratin production. The experimentally induced keratin-producing cysts were histologically similar to human odontogenic keratocysts.

Epithelial-lined cysts of the jaws fall into two main classes. The first category is derived from epithelial cells that have completed their biologic function in tooth formation and is usually associated with inflammation. They are lined by nonkeratinizing epithelium with discontinuities and various signs of degeneration. The second group is derived from epithelium that has not contributed to normal tooth structure. Such cysts are usually associated with an active keratinizing epithelium and are free of inflammatory cell infiltrates. Soskolne et al.[69] described a method for the experimental production of keratinizing cysts within rat mandibles by implantation of autogenous gingival epithelial grafts. In this study, 30 male albino rats, 140 to 185 g in weight, were anesthetized with pentobarbital sodium. A cavity was drilled into the exposed buccal surface of the mandible opposite the distal root of the first molar after reflecting the covering gingiva. The cavity was syringed with saline and plugged with a cotton pellet to control bleeding. The buccal aspect of the interdental papilla between the upper teeth was swabbed with 70% ethanol, washed with saline, and rinsed in 1% penicillin solution. Removal of the cotton pellet from the holes prepared in the mandible was followed by insertion of the incised papilla. Overlying tissues were sutured in place. The animals were sacrificed and the sites examined histologically after 0 to 45 days.

By the 8th day, 27 of the 28 animals had cysts with a lumen that was generally filled with orthokeratin. Cysts from the older time groups contained disintegrating keratin and free nuclear debris opposite areas of cyst wall where the lining was thin or completely absent. Although the histologic features of the experimentally produced cysts did not fulfill all criteria for odontogenic cysts, they had a number of features which resemble keratocysts, notably, keratinizing epithelium, no inflammatory cell infiltrate in the subepithelial connective tissue, and a well-delineated epithelial basal cell layer. The signs of degeneration and the discontinuities in the epithelial lining of the more mature cysts were reminiscent, however, of nonkeratinizing cysts.

A model for the development of follicular cysts wherein epithelial-lined structures about the crowns of murine tooth germs form in heterotropic sites has been developed by Riviere and Sabet.[70] Maxillary first molar tooth germs from 7-day-old mice of either sex were used as grafts and 2- to 4-month-old adult males as recipients. The grafts were placed in the fourth mammary fat pads that were surgically exposed and closed over with wound clips in the skin after graft emplacement. Donor mice were of the BBA/2J inbred strain, and recipients were either of the same strain (syngeneic transplant) or of the Balb/c inbred strain (allogeneic graft). Fat pads containing the transplanted tooth germs were removed 1, 2, and 3 weeks after grafting. They were processed histologically and read microscopically.

Cysts formed in every instance where the graft retained its viability and developed at about the same rate in all animals over the 3-week period. The histologic appearance coincided with that of human follicular cysts, but lacked keratinization and unusual cell types, presumably because of the nature of the implantation site. Spaces could be detected in the follicular epithelium prior to occult cyst formation. Cysts were found only in association with the developing crown, as is typical of the human counterpart. Only rudimentary cyst formation confined to the 1st week occurred in those teeth that were eventually destroyed by the host immune system. A viable undamaged follicular epithelium was a prerequisite for cyst formation. The stimulus for cyst production may have been a curtailment of blood supply, but cyst completion was dependent upon establishment of the graft in a new environment that permitted revascularization. Follicular cyst formation appears to be linked to an unsuccessful eruption process and to an innate susceptibility of the dental epithelium to cyst development.

Dentigerous cysts are thought to arise after completion of crown formation by accumulation of fluid either within the layers of reduced enamel epithelium or between the epithelium and enamel surface. Some hold that such cysts develop outside the follicle and are penetrated by an erupting tooth; others believe that some dentigerous cysts originate as periapical cysts on deciduous teeth into which permanent tooth crowns intrude. There have also been suggestions that some dentigerous cysts begin their formation by degeneration within the stellate reticulum during enamel apposition.

Al-Talabani and Smith[71] investigated the pathogenesis of dentigerous cysts with special reference to the frequency of enamel hypoplasia in the involved teeth. Maxillary second molar tooth germs from 2-day-old and 5-day-old hamsters were transplanted into the cheek pouches of 5- to 7-week-old syngeneic animals; 41 animals received a single tooth germ from 5-day-old hamsters in the left pouch. Recipient animals were killed in groups of four to seven at 4 days and at 1 to 24 weeks after transplantation. Another 27 hamsters received a single tooth germ from 2-day-old animals in the left pouch and were killed in groups of 8 to 10 at 2, 6, and 24 weeks after transplantation. Recovered transplants, together with surrounding host tissues, were fixed in 10% buffered formalin, demineralized, and double-embedded in nitrocellulose and paraffin wax. Serial sections were stained with hematoxylin and eosin and examined microscopically.

Cysts lined by epithelium were found frequently in association with tooth germ isografts in the cheek pouches. They developed from odontogenic epithelium in close relation to the crowns of the involved teeth. Cysts in tooth germ isografts from 5-day-old animals began to form shortly after transplantation as a result of enamel organ degeneration. The pathologic process that initiated cystic degeneration in the enamel organ in this instance was accompanied by degeneration of the ameloblasts and enamel hypoplasia. When tooth germs from 2-day-old hamsters were transplanted, cystic spaces developed only after completion of enamel formation as a result of separation between cells of the reduced enamel epithelium. Enamel hypoplasia was not a conspicuous feature. The experimental data suggest that there may be at least two types of dentigerous cysts, perhaps with different causes, arising at different stages of tooth development.

Burstone and Levy[72] produced experimental apical granulomas in 20 golden hamsters (*Cricetus auratus*) by fracturing the lower left first molar with a small rongeur forceps. Care was taken to avoid damage to the surrounding soft tissues. The animals were fed a diet of dog chow and water *ad libitum* and sacrificed at intervals between 3 and 186 days. The jaws were fixed in formalin, decalcified in formic acid, embedded in paraffin, sectioned serially, and stained with hematoxylin and eosin.

Microscopically, a periapical inflammatory response was noted after 8 days. The entire pulp was necrotic and the periapical region lightly infiltrated with neutrophils. After 23 days, granulomatous lesions were located at the apices of the fractured teeth with microscopic foci of fibroblastic and capillary proliferation; 42 days after fracture, the granulomatous lesions were diffusely infiltrated with chronic inflammatory cells that extended into the marrow spaces. After 72 days, the granulomas were walled off from the adjacent marrow by fibrous connective tissues. Small nests of degenerating inflammatory cells were located near the center. By 102 days, the granulomas consisted of a well-defined mass of young connective tissue surrounding a core of acute and chronic inflammatory cells. Involvement of the marrow spaces was extensive. At 186 days, the apical lesions had coalesced and filled the entire marrow space adjacent to the inferior alveolar nerve. Fibroblasts and round cells predominated. Many of the cells showed hydropic degeneration. Epithelial proliferation and cyst formation were found in several of the older lesions.

Proliferation of epithelium within apical granulomas appears to be a necessary precursor to cyst formation. Such proliferation has usually been attributed to a reaction to inflammation. Summers and Papadimitriou[73] studied the ultrastructure of apical granulomas to establish whether the epithelial proliferation is related to a phagocytic function. Small sections of human periapical granulomas containing proliferating epithelium were incubated with a radiopaque marker (10% Thorotrast) and tissue culture medium (modified Hanks) at 37°C for 30 min. They were then fixed in 2.5% buffered glutaraldehyde, postfixed in buffered osmium tetroxide, and embedded in Araldite. Sections were stained with lead citrate and examined in an electron microscope. Ultrastructurally, the apical granulomas were comprised of squamous epithelium, connective tissue on which the epithelium rested, and inflammatory cells. After challenge with Thorotrast, vacuoles containing electron dense matter were seen in the proliferating epithelial cells. These cells appear to have the property of microphagocytosis that endows them with the capacity to eliminate cellular debris and to participate in cyst formation.

EXPERIMENTAL SALIVARY MUCOCELES

Lesions similar to human salivary gland mucoceles have been created experimentally in mice, rats, and cats. Bhaskar et al.[74] exposed the submandibular group of salivary glands of five female mice, age 6 months, and of six female rats, age 3 months, by skin incision. The right submaxillary excretory duct was dissected out and severed. Animals were sacrificed at intervals from 1 to 9 days postoperatively. At necropsy, the glands were studied grossly, removed, fixed in 10% formalin, embedded in paraffin, cut serially, and stained with hematoxylin and eosin, Mayer's mucicarmine, silver impregnation, Mallory's connective tissue stain, and van Gieson's stain.

The submandibular group of glands on the experimental side appeared much larger than those on the control side on gross comparison. In five of the rats and three of the mice, the superficial surface had circumscribed either translucent swellings or cystic cavities of varying size filled with a clear viscous fluid resembling mucus. In one rat and two mice, the submandibular area contained fistulae and grayish-yellow necrotic areas. One day after duct severance, the interglandular connective tissue and lymph nodes evidenced ill-defined areas of intercellular basophilic vacuolated material in which numerous cells were embedded. The material was mucicarmine positive. Connective tissue surrounding and separating the mucus pools showed varying degrees of neutrophilic infiltration; 2 to 9 days after duct severance, typical mucoceles developed in the mice and rats comparable to those of man. The cyst walls were not lined by epithelium. In sections stained with silver, argyrophil fibers extended to the cystic cavity. Mucoceles apparently result from trauma that permits escape of saliva into the adjacent tissues.

Chaudry et al.[75] tested the effects of excretory duct ligation, severance, and compression on mucocele formation in rats. In that test, 45 male white rats, 4 months of age, were divided into three groups. Each contained an equal number of animals of comparable weight. The rats were anesthetized with Nembutal® and maintained with ether while the right submandibular gland was exposed surgically through a ventral midline incision and the ducts separated from the fascia. In Group I the duct was ligated near exit from the gland; in Group II the duct was cut at the same level; in Group III the duct was pinched with a hemostat. Animals in Groups I and II were sacrificed at 3, 7, and 14 days; those in Group III, at 1, 3, and 7 days. At necropsy, the submandibular glands were removed, fixed in 10% formalin, and processed in routine fashion for microscopic examination of hematoxylin and eosin-stained slides.

In the Group I animals, duct ligation for 3 and 7 days resulted in dilatation of the ductules, rupture of the ductules, adenitis, necrosis, atrophy of the glandular parenchyma, and interstitial fibrosis. By 14 days, the gland was shriveled to a fibrous mass. In Group II, there was pronounced glandular enlargement at 3 and 7 days. The glands were soft, smooth, shiny, and evidenced one or more areas of mucus accumulation when cut. Microscopically, they contained mucus intermixed with granulation tissue or mucus surrounded by granulation tissue not unlike human mucoceles. At 14 days, the glands were mostly atrophic and replaced by fibrous connective tissue. The changes in the Group III animals at 1 and 3 days were similar to those in Group II. Half had mucocelelike lesions by this time. The vast majority of mucoceles were apparently formed by the escape of mucus into the surrounding tissues from a mechanically injured excretory duct. The mucus either diffused into the tissue or formed a discrete pool surrounded by a wall of connective or granulation tissue with varying degrees of maturity. All subsequent changes were secondary to the accumulation of mucus and the inflammatory reaction. Complete obstruction did not produce mucoceles, as evidenced by their absence in the rats with the ligated excretory ducts. Severance or pinching of the ducts permitted production of mucocelelike lesions.

Harrison and Garrett[76] found that duct ligation will induce mucoceles in the sublingual gland of cats, but not in the submandibular and parotid glands. All three major salivary gland ducts were ligated in mature cats for periods ranging from 1 day to 1 year. Two 000 braided silk sutures were used for each ligation. Because the sublingual and submandibular gland ducts run together in the cat, they were ligated unilaterally anterior to the lingual nerve to avoid damaging the chorda tympani.

Extravasated mucus was found in all sublingual glands up to 20 days after ligation. Extravasation mucoceles developed in association with 12 of 27 sublingual glands between 7 days and 1 year after ligation. Enzyme histochemistry revealed that the extravasated mucus induced a migration of macrophages and a fibroblastic reaction. When these reactions were sufficiently intense in the days following ligation, continued spread of extravasation of mucus did not occur. When the extravasation of mucus was too great to be contained, a mucocele formed in which a balance developed between mucus extravasation and removal. Microscopic examination of the mucoceles disclosed that they do not have an epithelial lining consisting of mucus from ruptured acini surrounded by spread-limiting connective tissue.

The interrelations between excretory duct obstruction, salivary mucoceles and sialadenitis were studied by Bhaskar et al.[77] The right submaxillary gland duct of 31 female mice, age 6 months, was surgically exposed, ligated, and the wound closed. Animals were sacrificed at intervals between 1 week and 9.5 months. The glands were fixed in 10% formalin, embedded in paraffin, sectioned serially, and stained with hematoxylin and eosin, van Gieson's stain, Mayer's mucicarmine, and erythrosin-safranin. Ductal obstruction by ligation produced an early acute and later chronic interstitial inflammation of the submaxillary gland. Following ligation, there was a degeneration of the secretory granules and vacuoles and change in configuration to cuboidal or polyhedral cells. Ligation led to an increase in number of ducts formed by ductal cells and nonfunctioning or partially active acinar cells. The inflammatory and degenerative changes in the mouse submaxillary gland induced by duct ligation were comparable to those following obstruction of the salivary glands in man. Duct ligation did not produce lesions resembling human mucoceles.

Leake et al.[78] detailed the ultrastructural changes in the rat parotid gland after ligation of Stensen's duct. The duct was ligated unilaterally in 18 mature Sprague-Dawley rats, and the parotid gland was examined by electron microscopy at intervals up to 90 days. The course ran from swelling and engorgement lasting 4 to 6 days to progressive atrophy. At 3 months, the number of identifiable acini was greatly decreased. Nuclei in the few remaining acinar cells were prominent because of the lack of secretory granules in the cytoplasm. Endoplasmic reticulum was severely disorganized and sometimes absent. Golgi apparatus was markedly decreased. Microvilli were truncated with blunt ends. Cell borders remained distinct with well-demarcated desmosomes. There was an increase in collagen and infiltration of some inflammatory cells in the areas formerly occupied by the acini.

Standish and Shafer[79] described the histologic changes in rat submaxillary and sublingual glands following ligation of the excretory ducts and/or arteries. Weanling male Sprague-Dawley rats were anesthetized with Nembutal® and the right and left submaxillary and sublingual glands were exposed by a midline incision. The arteries and ducts were separated by blunt dissection and a 0000 black silk suture tied around the duct in one group, the artery in a second group, and both the duct and artery in a third group. The wounds were closed and the animals fed a commercial laboratory stock diet and water *ad libitum*. Animals were sacrified between 1 day and 20 weeks following ligation. The salivary glands were removed en masse and fixed in Helley's solution, processed, sectioned, and stained with hematoxylin and eosin and special stains when indicated. Controls consisted of the opposite unligated glands.

Duct ligation alone resulted in progressive atrophy of the acinar cells and reversion to a resting or inactive state. Acute ligation did not result in the formation of retention cysts. Artery ligation produced typical infarcts, squamous metaplasia of parts of the duct system, and repair of the infarct by regeneration of acinar tissue. Combined duct and artery ligation produced features associated with each. Infarction was more complete, and there was a failure of acinar regeneration.

Bhaskar et al.[80] studied regeneration of the submaxillary gland of rabbits following experimentally induced obstructive atrophy. In this study, 71 male adult New Zealand rabbits were divided into three groups. Group I comprised 10 unoperated controls. Group II contained 25 animals in which the major submaxillary gland ducts were ligated by an intraoral surgical approach. Five rabbits in the group were killed 3 days later and the glands recovered and used as experimental controls. In the others, the ligated duct was reexposed and cut proximal to the ligature to eliminate the obstruction. They were killed in pairs 1 to 54 days later and the glands removed for study. Group III included 36 animals with a ligated major submaxillary gland duct. Six were killed 7 days following ligation. In the others, duct obstruction was relieved after 7 days by surgical intervention. They were killed in pairs after 1 to 49 days and the glands retrieved. The glands were fixed in 10% formalin, bisected parallel to the long axis, embedded in paraffin, cut serially, and stained with hematoxylin and eosin.

Obstruction of the submaxillary gland duct produced atrophic changes that were well advanced by 3 days and quite severe by 7 days. At 3 days, the glands were reduced in size and the acinar cells were small, having lost basophilia, vacuoles, and secretory granules. The trophochrome cells had disappeared and the secretory tubules were small and lined with cells of reduced height. The inter- and intralobular ducts were dilated. By 7 days, the acini had disappeared and were replaced by islands of small polyhedral cells and connective tissue infiltrated with plasma cells, lymphocytes, and eosinophils. Following removal of the duct obstruction, the glands evidenced regeneration that began within a day and continued for about a week. There was a reappearance of the acinar cells, trophochrome cells, and secretory material. Regeneration was faster in glands ligated for 3 days than in those obstructed 7 days. After 1 week, the unobstructed glands began to form mucoceles in the surrounding connective tissue. Ducts again became obstructed followed by acinar atrophy.

EXPERIMENTAL ORAL ARTERIOPATHIES

Castelli et al.[81] investigated the effect of experimentally contrived interruption of the alveolar arterial blood flow on mandibular collateral circulation and dental tissues. Nine young adult rhesus monkeys of both sexes with a permanent dentition were used in the study. Under general anesthesia, the parotid gland and masseter muscle on one side were exposed surgically through a vertical cutaneous incision. The parotid was packed posteriorly and the masseter incised vertically to reach the lateral border of the ramus. Drilling in the area of the mandibular ramus exposed the inferior alveolar artery. The vessel was damaged by cauterization, and the wound closed. A daily i.m. injection of penicillin-streptomycin was given 4 days postoperatively to prevent infection. The animals were killed at intervals between 1 day and 9 weeks after surgery. Both external carotid arteries were immediately cannulated and a radiopaque material infused into the facial area for angiogram study. All heads were fixed in 10% buffered formalin and processed for microscopic examination of the teeth, periodontal membranes, and alveolar bone.

Interruption of circulation through the inferior alveolar artery was followed by the establishment of a fast retrograde blood flow through the vessel. The mental artery and the mandibular branch of the sublingual artery were the main vessels to contribute to this flow. This proved to be decisive in sustaining the nutritional needs of the mandibular body and in preventing bone infarction immediately after the vessel was cauterized. No histopathologic changes were found in the experimental hemimandibles. Odontoblasts in the pulps of the molar teeth suffered temporary regressive changes following cautery of the inferior alveolar artery.

Medial arteriosclerosis can be produced experimentally in the rat by creation of an obstructive nephropathy that leads to renal hypertension. The ensuing vascular lesions resemble in histologic detail those typical of Monckeberg's medial sclerosis in man. Dreizen et al.[82] induced this arteriopathy by restricting rats to a diet of skimmed milk powder and tap water. Although confined most often to the arch and thoracic segments of the aorta, generalized vascular involvement was found in 14 animals. The 14 male Sprague-Dawley rats had been started on the experimental regimen at 6 weeks of age and continued until expiration. They were matched with a control series in sex, number, and age, and maintained on a stock diet until sacrificed upon death of the paired test animal. At necropsy, the heads were fixed in 10% buffered formalin, decalcified in nitric acid, tissue blocked, processed for paraffin-embedding, sectioned, and stained with hematoxylin and eosin. Selected sections were stained by the von Kossa method.

Medial arteriosclerosis was found in the large branches of the lingual and buccal arteries. Only branches with a comparatively thick tunica media were affected; thin walled arteries and arterioles were spared. In the afflicted vessels, the normal tunica media pattern of parallel elastic lamellae alternating with layers of smooth muscle was replaced by patches of degeneration and necrosis involving both structural elements. Elastic lamellae were split and fragmented, and the smooth muscle was replaced by sclerotic and necrotic debris that contained accumulations of connective tissue ground substance, collagen scars, and deposits of calcium salts. There were no detectable structural alterations in the oral tissues supplied by the sclerosed arteries. Arterial medial calcification does not lead to occlusion of the lumen or rupture of the vessel wall and has few, if any, clinical sequelae in the mouth (Figure 2).

EXPERIMENTAL ORAL DRUG REACTIONS

Bhussry and Rao[83] examined the reaction of the rat oral mucosa to sodium diphenylhydantoin. Five litters, each containing ten newborn albino rats, were used in the experiments. Beginning at 3 days of age, eight rats in each litter were given an i.p. injection of 0.04 m*l* suspension of 100 mg/kg body weight dilantin sodium in isotonic saline every 3rd day. The two other rats in each litter were injected with 0.04 m*l* isotonic saline i.p. and served as controls. Test rats received from one to five doses of the drug. Two test animals from each litter were killed 6 days after administration of

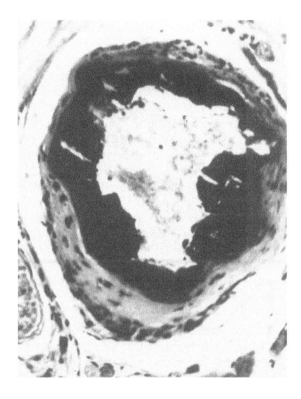

FIGURE 2.   Branch of lingual artery from rat with experimentally induced medial arteriosclerosis showing deposits of
calcium in vessel wall. (Hematoxylin and eosin stain, Magnification × 100.) (From Dreizen, S., Stone, R. E., and Stahl, S.
S., *Arch. Oral Biol.*, 8, 187, 1963. With permission.)

the last dose, along with an appropriate control. The oral tissues were processed for
microscopic examination and stained with hematoxylin and eosin, Masson's trichrome,
PAS, and Mowry's procedure for acid mucopolysaccharides.

The initial histologic change in the oral mucosa of rats given one dose of drug was
increased mitotic activity in the basal cells of the epithelium. Intercellular edema within
the epithelium produced prominent spongiosis and acanthosis, and an apparent disturbance in cornification resulting in parakeratosis. Epithelial hyperplasia was common in animals receiving four doses of dilantin sodium. The new cells were devoid of
nuclear detail, and the rete pegs were elongated and split. Large numbers of discrete
basophilic keratohyaline granules were diffusely distributed throughout the epithelium
in association with hyperkeratosis. Elevated fibroblastic activity was accompanied by
increased collagen fiber formation. Histochemical staining reactions suggested a substitution of glycoproteins for mucopolysaccharides in the epithelial cell layers, and an
increase in glycoproteins and mucopolysaccharide-staining material in the connective
tissues.

Staple et al.[84] introduced the stump-tailed macaque (*Macaca arctoides*) as an experimental model for diphenylhydantoin gingival hyperplasia. Relegation of this species
to a soft diet without oral hygiene results in rapid deposition of dental plaques with
associated gingivitis. Four adult female stump-tailed macaques, average weight 7.3 kg,
age unknown with all third molars erupted, were followed through seven experimental
phases consisting of: (1) oral hygiene to reduce plaque deposition and gingivitis to
baseline values, (2) experimental gingivitis from plaque accumulation in absence of

oral hygiene, (3) repeat Phase I while animal is trained to take medication using the vehicle for diphenylhydantoin, (4) repeat Phase II with addition of diphenylhydantoin administration, (5) plaque effect determined during elimination of diphenylhydantoin while vehicle given, (6) plaque effect alone studied, (7) hygiene measures resumed to restore oral conditions to original baseline.

Diphenylhydantoin was administered as an oral suspension at an initial dose of 20 mg/kg once or twice daily. This was titrated to maintain plasma levels of 10 to 20 μg/ ml. The extent of gingival hyperplasia was measured from dental casts made from alginate impressions of both arches taken at intervals throughout the study. Gingival hyperplasia was quantitated from the casts by planimetry of the exposed facial surfaces of selected teeth in lateral view photographs. The expansion of the interdental gingival papillae and marginal gingiva by hyperplasia resulted in progressive decreases in amount of exposed tooth surface. Buccal gingival biopsies were taken at intervals, fixed in buffered formalin, embedded in methacrylate or paraffin, sectioned, and stained with hematoxylin and eosin or toluidine blue.

No toxic actions were observed when the plasma level was 10 to 20 μg DPH/ml. Slight loss of body weight occurred at higher levels. When the animals were maintained on a soft diet without oral hygiene at plasma levels of 10 to 20 μg DPH/ml, gingival hyperplasia became evident in those sites where the plaque and gingival indices were highest. Dental ligatures enhanced the development of plaque-mediated gingivitis and of diphenylhydantoin-mediated gingival enlargement. The interdental papillae were the first to enlarge mesiodistally and buccolingually. Clinically and histologically, the diphenylhydantoin gingival enlargement in the monkeys resembled that in human users of the drug. Gingival enlargement regressed after the vehicle (Phenylcyclidine) was substituted for diphenylhydantoin and the animals were continued on the regimen that maintained plaque deposition and gingivitis. Regression began before the plasma and saliva levels of diphylhydantoin reached zero. Diphenylhydantoin-mediated gingival enlargement was completely reversible within 26 weeks after drug cessation.

The tetracycline antibiotics have been shown to concentrate in growing bones and teeth. Children treated with such antibiotics during the period of tooth formation develop irreversible yellow to yellow-brown discoloration of the teeth. Owen[85] studied the capacity of different tetracyclines to localize in the teeth of young dogs. A litter of crossbred retriever dogs, 7 weeks old, were dosed for 1 month with therapeutic amounts of various tetracyclines. Each dog received either tetracycline, chlortetracycline, oxytetracycline, or dimethylchlortetracycline. Another litter received very high doses of the same drugs.

High doses of chlortetracycline inhibited bone growth. In most cases, the primary teeth became yellow and fluoresced yellow in UV light. Permanent teeth examined soon after eruption appeared very yellow in dogs given dimethylchlortetracycline, less yellow in dogs that received tetracycline or chlortetracycline, and white or dull-white in dogs treated with oxytetracycline. Under UV light, all teeth fluoresced yellow or orange-yellow. Examination of ground sections for fluorescence revealed that the tetracyclines were incorporated in both the enamel and dentin at the time of tooth formation.

The oral mucous membranes are frequently affected by undesirable side reactions in patients receiving cancer chemotherapy. Mucositis, glossitis, and ulcerative stomatitis are encountered after the administration of antifolic derivatives, antipurines, antipyrimidines, and other oncolytic agents. The administration of such cytotoxic drugs produces deficiencies in cellular metabolism that interfere with cell growth, cell maintenance, and cell replication.

Monto et al.[86] determined the effects of cancer chemotherapy agents on cells of the buccal mucosa obtained by scraping before, during and after treatment of patients with nitrogen mustard, 6-mercaptopurine, piperazine, methotrexate, or 5-fluorouracil. Surface areas of the cytoplasm and nucleus were determined by measuring the two greatest diameters. The measurements and other nuclear and cytoplasmic properties were compared with cells from normal controls and from pretreatment and posttreatment smears. Piperazine increased the cytoplasmic and decreased the nuclear diameter; 6-mercaptopurine decreased both diameters; methotrexate increased both diameters; 5-fluorouracil decreased the cytoplasmic and increased the nuclear diameter. The changes produced by the cancer chemotherapeutic agents were similar to those noted in clinical deficiencies of folic acid, purines, and pyrimidines, indicating that disturbed nucleoprotein synthesis was the common underlying denominator.

Actinomycin D is a cytostatic antibiotic that interferes with the biosynthesis of proteins and messenger RNA. Actinomycin D is presently used to treat some forms of cancer because it combines with DNA to block DNA-dependent RNA polymerase activity. Ghee et al.[87] studied the salivary glands of rats given a single injection of a sublethal dose of Actinomycin D for cytologic changes. In this study, 21 adult male Sprague-Dawley rats were injected i.p. with 75 μg Actinomycin D dissolved in 0.5 m*l* saline. Seven rats served as controls and received the same volume of saline alone. Animals were fed *ad libitum* on regular laboratory chow. Three experimental and one control rat were sacrificed between the first and 21st day after drug administration. At necropsy, the right parotid and submandibular glands were dissected out and weighed. Small pieces were fixed in Bouin's and in Zenker's formol, double-embedded in Parloidin-paraffin, sectioned, and stained with hematoxylin and eosin. Other sections were stained for nucleic acid with methyl green-pyronin, for RNA and cytoplasmic granules with PAS-Azure II, for parotid gland cytology with aldehyde fuchsin-Masson, and for submandibular gland cytology with mucicarmine-hematoxylin and PAS-Alcian blue. The ratio of number of nucleoli to nuclei was determined to obtain a quantitative estimate of the nucleolar changes in the secretory cells. The parotid and submandibular glands underwent significant weight losses during the first 7 to 14 study days. Histologically, the number and size of nuclei in the secretory cells decreased rapidly within 24 hr after injection. Advanced stages of nuclear pyknosis with concomitant decrease in cytoplasmic basophilia and granule content were noticeable by the third study day. Vacuolization and rarefaction of the cytoplasm was apparent by day 7. Nuclear and nucleolar recovery began on day 10 with complete cytoplasmic recovery evidenced by day 21. Cytologic damage in the parotid was greater than in the submandibular acini.

Adkins[88] delineated the ultrastructural Actinomycin D induced changes in the acinar cells of the rat submandibular gland. In this study, 16 Wistar albino rats weighing approximately 250 g were randomly divided into 4 groups each containing 1 control and 3 experimental animals. The latter were given a single i.p. injection of 75 μg Actinomycin D per rat. Groups of four were killed by decapitation at intervals between 1 and 9 days after injection. No solid food was provided in the 24 hr immediately preceding except for the very last hour. This was done to insure that all glands would be in a similar state of function at the time of death. The glands were fixed in glutaraldehyde and processed for electron microscopic examination.

There were no detectable structural changes in the submandibular gland in the first 24 hr. At 3 and 6 days following injection of Actinomycin D, the arrangement and shape of the cells and the location and morphology of some of the cytoplasmic structures and organelles were aberrant. Ballooning degeneration in many mitochondria and development of numerous vesicles, cytolysosomes, and myelin figures were prominent features. By the 9th day postinjection, there was evidence of early resumption of synthesis, and secretion granules were increasing in number. Whorled arrangement of endoplasmic reticulum was a distinctive feature in the treated cells.

Single injections of the antitumor agent vincristine produce immediate, varied, and dose-dependent effects on the odontoblastic population of the rat incisor. An initial mitotic arrest of cells in the germinative part of the pulp, and swelling and necrosis of some odontoblasts, are followed by severely altered cell morphology and irregular dentin production in some parts of the incisor. Stene[89] examined the effect of a single injection of vincristine after 1, 2, and 3 weeks to determine whether the derangements are permanent and whether there are any late manifestations of vincristine toxicity in the continuously developing rat incisor. Vincristine was administered i.v. to 60 rats in doses of 0.1, 0.3, 0.5, and 0.7 mg/kg. Histomorphologic study of the dentin and odontoblasts in the maxillary incisors disclosed moderate to severe disturbances in dentin production in parts of the incisor. In some, the effect seemed to be reversible since normal-appearing dentin was found pulpal to the dentinal derangements after 2 or 3 weeks. The long-term distribution and severity of the vincristine-induced dentinal lesions were similar to those occurring shortly after injection of the same dose of drug.

In response to evidence suggesting that salicylates may be teratogenic, Yen and Shaw[90] elicited the effects of repeated ingestion of large quantities of acetylsalicylic acid on mandibular bone growth and on dentin apposition in young monkeys. Three rhesus males weighing 3 to 3.4 kg with mixed dentition were maintained on a commercial pelleted diet throughout the experiment. All were given i.v. injections of lead acetate (3 mg/kg body weight) as an intravital stain every 7 days for 42 days. Two of the monkeys also received an oral dose of fruit-flavored acetylsalicylic acid at a level of 325 mg/kg body weight daily for 4 days beginning on the day of the second lead acetate injection. The third monkey received saline and served as a control. Each was killed by an overdose of phenobarbital 4 hr after the last injection of lead acetate. Sections of the zygomatic arch and mandibular ramus were removed, fixed in 10% formalin, decalcified, and processed for histologic examination. Unerupted permanent mandibular canines and second molars were also removed and decalcified. During decalcification the acid solvent was saturated with hydrogen sulfide gas to form black lead sulfide at the mineralizing sites. The teeth were then prepared for histologic study.

Four doses of acetylsalicylic acid inhibited membranous bone growth and remodeling of the Haversian canal system in the mandible for 16 to 18 days with incomplete recovery during the experimental period. In contrast to the inhibition of osteogenesis of the membranous bone, dentinogenesis was unimpaired. The inhibitory effect of salicylates on the synthesis of acid sulphomucopolysaccharides, an important component of ground substance, was offered as the presumptive mechanism for salicylate-induced skeletal malformation.

EXPERIMENTAL NERVE OR MUSCLE TRANSECTION

Butcher and Taylor[91] examined the effects of denervation and ischemia on the teeth of the monkey. This study was designed to define the consequences of cutting the inferior alveolar nerve during resection of the mandible and of interfering with nerves and blood vessels during oral surgery procedures like the Caldwell-Luc operation. Rhesus monkeys with a permanent and with a deciduous dentition were used in these experiments. Animals were anesthetized with i.p. sodium pentothal and the inferior alveolar nerve exposed by an extraoral approach. In the initial experiments, ligatures were tied at each end of the exposed nerve and a transection made between ligatures. In later experiments, the nerve was cut close to the mandibular foramen and the proximal end sutured to the parotid gland fascia or to the external pterygoid muscle. To deprive the teeth of vascularity for selected intervals, methods such as ligating the inferior alveolar and mental arteries and retraction of the incisors into their sockets by forced tension were used. To investigate the influence of ischemia and denervation on the contents of the pulp, either the apex, apical third, or entire pulp was removed. The operations were usually performed on the upper central incisors by an external approach. All animals were sacrified and the tissues examined histologically.

Tooth development and maintenance of tooth structure are not dependent on innervation since they continued to occur without a nerve supply. Unless the blood vessel disturbance was in the apical region of the tooth, ischemia severe enough to cause permanent tooth injury could seldom be established. Pulps of teeth with large apical foramina were more easily injured by retraction than the pulps of teeth with small foramina. While odontoblasts and most pulp cells degenerated upon retraction, they were replaced from the compressed tissues at the apex when the pressure was released.

The effect of glossopharyngeal nerve transection on the circumvallate papillae of the rat was reported by Guth.[92] In this study, 32 female Osborne-Mendel rats weighing between 100 and 150 g were anesthetized with chloral hydrate and the base of the skull exposed to visualize and transect the glossopharyngeal nerve on each side as they crossed the tympanic bulla. Cuts were made distal to the petrosal ganglion. Control groups consisted of five sham-operated and three unoperated animals. Groups of animals were anesthetized each day for removal of the posterior part of the tongue. Specimens were rapidly fixed and processed for histologic examination. The number of taste buds in each circumvallate papilla was determined from serial sections.

During the first postoperative day, there were no significant changes in the number of taste buds. The number then declined rapidly for 3 days and slowly after the 4th day. Virtually all taste buds were gone by the 7th day. Taste buds were lost by desquamation from the epithelial surface and replaced by stratified squamous epithelium. The taste buds showed degenerative changes prior to extinction. Glossopharyngeal nerve resection also produced an atrophy of the epithelium covering the circumvallate papillae and of the entire lingual mucosa. The morphologic integrity of the taste buds is, thus, highly dependent on an intact nerve supply.

Boyd et al.[93] evaluated the effects of loss of muscle tension on the coronoid process of the mandible in the presence of an intact blood supply to the bone. Ten guinea pigs were anesthetized with i.p. sodium pentobarbital and lidocaine hydrochloride in the region of surgery. The temporalis muscle was exposed by an incision made over the left ear through the tela subcutanea and fascia. The entire origin of the muscle was resected from the skull, and the freed muscle was allowed to roll up on itself before the incision was closed. The unoperated side served as a control. All animals were sacrificed 80 days after surgery. Mandibles were removed, cleaned of tissue, and gross comparisons made of the coronoid process between the operated and unoperated sides and between the operated animals and normal guinea pigs of the same age and weight.

Removal of the origin of the temporalis muscle from the skull to eliminate muscle tension on the coronoid process, while still maintaining an intact blood supply to the mandible, did not change either the size or shape of the coronoid process in the majority of animals. Removal of masticatory muscles in previous studies where the blood supply to the bone was eliminated caused a decrease in size of the region of insertion on the coronoid process. This study demonstrated that the effect is due to loss of blood supply and not to loss of muscle tension.

EXPERIMENTAL DRUG-INDUCED CLEFT PALATES

About 25% of all cleft palate malformations are genetic in origin. In the majority of others, both environment and heredity participate in the etiology. Among the environmental agents are drugs, nutritional deficiencies, hypervitaminosis A, and ionizing radiation. Drugs most culpable in this respect are steroids and cytotoxic agents. Chaudry et al.[94] determined the effect of cortisone on the incidence of induced and spontaneous cleft palates, fetal resorption, and birth weights in A/Jax mice when the mothers were injected on various days of gestation. A comparison was also made between the effects of i.m. and i.p. injection. A/Jax mice, 3 to 4 months old, were fed Wayne Lab chow and water *ad libitum* and kept in a room with an automatic switch for alternating 12-hr periods of light and darkness. Three primiparous females were caged with one male during darkness and examined for copulation plugs during the next light period. Females with vaginal plugs were separated and caged alone. Conception was arbitrarily assumed to have occurred at the midpoint of darkness, and gestation was computed from that time.

Pregnant females were assigned at random to one of four experimental groups. They were given a single i.m. injection of 10 mg cortisone on either day 11.5, 12.5, 13.5, or 14.5 of gestation. Another group was given a single i.p. dose of 10 mg cortisone on day 12.5. A control group was subjected only to needle trauma. On day 18.5 postconception, the animals were anesthetized with sodium pentobarbital and the fetuses delivered by Caesarean section. Numbers of viable resorbed fetuses were recorded. Fetuses were fixed in 10% formalin and examined for clefting of the palate the next day.

Induced cleft palates involved only the hard and soft palates; spontaneous cleft palates were associated with unilateral and bilateral cleft lips. The single injection of cortisone on days 11.5, 12.5, and 13.5 of gestation yielded 85.0, 92.5 and 83.0% of induced cleft palates, respectively. The teratogenic effect of the drug was greatly diminished when given on day 14.5. Incidence of spontaneous cleft palates increased as the day of injection advanced. I.p. administered cortisone produced a significantly lower incidence of cleft palate than did i.m. cortisone. The fetal birth weights were substantially lower in the cortisone-treated animals than in the controls.

Blaustein et al.[95] compared the cleft palate inducing activity of corticosterone, ACTH, epinephrine, cortisone, and hydroxycortisone in mice under standardized and reproducible conditions. Random-bred female CD-1 mice received the 5th or 6th day of pregnancy were housed three to five in a cage and fed standard breeder mouse chow and water *ad libitum.* The five drugs were suspended in the same vehicle and injected either i.m., i.p., or s.c. on days 11 to 14 of pregnancy. The volume of each injection was 0.1 mℓ. One group was not injected and served as controls. Cortisone, corticosterone, and hydroxycortisone were administered at a dose of 2.5 mg each day for 4 days, ACTH at a level of 10 units per day for 4 days, and epinephrine in an amount of 0.1 mg per day for 4 days. The mice were killed on day 17. Each uterus removed was examined for absorption sites and dead fetuses. Live fetuses were removed, fixed in Bouin's, and examined for cleft palates. Cortisone and hydroxycortisone greatly increased cleft palate incidence regardless of the injection route. Corticosterone was not teratogenic by the i.p. route. ACTH and epinephrine did not induce cleft palates under the conditions of this study.

Metah et al.[96] developed a model for precipitating 100% cleft palates in mice. Five groups of pregnant Swiss mice were given a single i.p. injection of either 5, 10, 20, 30, or 40 mg/kg body weight of cyclophosphamide, an antitumor alkylating agent on one of days 7.5 to 13.5 of pregnancy, inclusive. On day 18.5, the uterine horns were incised following laparotomy and the number of fetuses, positions, and resorptions were counted. The fetuses were examined for abnormalities of the cranium, extremities, and body. Cyclophosphamide caused cleft palate formation in all fetuses when given to the mother at a concentration of 30 mg/kg body weight on day 11.5 of gestation. When the same amount was given 1 day later, the incidence dropped to 90% and fell sharply thereafter.

EXPERIMENTAL IRRADIATION OF THE ORAL STRUCTURES

Progressive necrosis of the mandible and maxilla with and without superimposed infection is often an undesirable sequel to radiation treatment of benign and malignant tumors in and about the oral cavity. The clinical manifestations of irradiation to the oral tissues were studied by Medak and Burnett[97] in six rhesus monkeys approximately 3 years of age. Each was irradiated daily to the same site on the right cheek with 600 R X-irradiation. One monkey received a total of 4200 R; two, 5000 R; and three, 6000 R. The exposure produced an ulcerative mucositis in the cheek and tongue that varied in severity with the intensity of the irradiation. These lesions healed within 4 to 6 weeks after irradiation. Following irradiation, there were two distinct types of reaction in the mandible and maxilla. The more severe was a radioosteitis represented by necrosis of the bone and gross infection with sloughing of the gingiva and loss of bone. The less severe was a radioosteonecrosis characterized by bone necrosis without infection.

Burstone[98] detailed the histologic changes in oral tissues of mice in response to continuous irradiation. In this study, 30 young adult C57 black mice were injected s.c. with a solution of carrier-free P^{32}. Dosages of 30, 20, and 10 μCi/g of body weight were used. Controls consisted of untreated mice and of mice injected with nonradioactive phosphoric acid in quantities similar to the radioactive preparation. All animals were killed 7 to 10 days after injection. The heads were split longitudinally, fixed in 10% formalin, decalcified, processed, sectioned, and stained with hematoxylin and eosin.

Radiophosphorus in amounts of 20 to 30 μCi/g of body weight produced extensive and destructive changes in the tongue and oral epithelium. Lingual epithelial cells, especially those in the stratum germinativum, underwent enlargement and exhibited hyperchromatic nuclei. There was a break in the continuity of the epithelium on the dorsum covered in part by a keratin layer that still resembled the filiform papillae in general outline. Foci of necrosis in the epithelium and underlying connective tissue were apparent 7 to 10 days after injection. Early radiation effects on the temperomandibular joint were marrow aplasia and swelling of the cartilage cells adjacent to the zone of erosion.

Liu et al.[99] developed a method for irradiating the everted lower lip of the rat and shielding the rest of the body that permitted examination of ultrastructural changes in the oral epithelium in response to X-irradiation. This avoided complications due to ulceration and modification of the mucosal response by total body irradiation. Adult male rats of the Simonsen strain were given a dose of 5000 R X-irradiation directed at the lower lip everted through a hole in a lead rubber cylinder shielding the head and body. Light and electron microscopic observations were made on specimens of lip oral mucosa from animals killed between 2 hr and 12 days after irradiation.

Widespread degenerative changes were noted in the basal cells as early as 2 hr after irradiation, increasing in degree up to 50 hr. Inflation of the outer nuclear envelope and rough endoplasmic reticulum with loss of ribosomes, swelling of mitochondria, and disarrangement of the cristae were evident at 2 hr, followed by swelling of the nucleus and cytoplasm at 6 hr, and by frank membrane breaks at 26 hr. Irreversible damage was noted in a small but growing minority of cells. Immediate mobilization of the Golgi-lysosomal system was manifested by increase in the size of the zone and subsequent autophagic activity. Beginning at 26 hr, signs of recovery were noted in the nuclear envelope, rough endoplasmic reticulum, and intercellular space. Mitotic activity was apparent by 50 hr, and recovery was virtually complete by 12 days.

Microscopic reflections of early radiation injury in the rat parotid gland were elucidated by Sholley et al.[100] Female Sprague-Dawley rats weighing between 170 and 230 g received either 1600 or 6400 R in a single exposure delivered at a rate of approximately 190 r/min. Trypan blue and colloidal carbon were used as tracers to detect abnormal vascular permeability. Trypan blue was administered in a dose of 0.2 m*l* of a 2% solution in saline per 100 g body weight. The tracers were injected into the lateral tail vein of the anesthetized rats. Some rats received both tracers; most received only one. Circulation time ranged between 10 and 90 min before sacrifice. Parotid glands were examined by light and electron microscopy after appropriate fixing and processing. For light microscopy, the glands were fixed in 10% formalin and prepared for paraffin sections stained with hematoxylin and eosin and with special stains. For electron microscopy, the glands were fixed in Karnowsky's fixative, postfixed in osmium tetraoxide, dehydrated in ethanol, and embedded in Epon®.

Acinar cells were the most sensitive glandular components showing morphologic evidence of radiation damage as early as 2 hr. The damage appeared either as focal cytoplasmic degeneration resembling large cytolysosomes or as necrosis of the acinar cells expressed by formation of unusual structures termed "light bodies." Damage became maximal by 2 days, followed by a 3-day period of cleanup and acinar atrophy. More extensive injury was produced by 6000 R than by 1600 R. By 4 days after exposure, glands subjected to the higher dose had a 47% weight loss. Loss of acini and decrease in acinar size were evidenced by extensive basal lamina redundancies attached to shrunken acini or lying free in the interstitium. Surviving acinar cells appeared relatively normal 8 days after exposure, although most contained basal lipid vacuoles and less than the normal complement of secretory granules. Neither dose produced significant alterations in the ducts or microvasculature.

Shafer[101] determined the effects of single and fractionated doses of X-irradiation on the histologic structure of the major salivary glands of the rat. Male albino Wistar rats weighing 100 to 400 g were exposed to a single dose of X-irradiation ranging from 480 to 3500 R, while another group was given fractionated doses ranging from 4500 to 8000 R. All were sacrificed 21 days after the last treatment. The salivary glands were dissected out *in toto* and fixed in Helley's solution. Paraffin-embedded sections were stained with hematoxylin and eosin, Mallory's trichrome, phosphotungstic acid-hematoxylin, and basic fuchsin-hematoxylin-light green. Rats exposed to the single doses of radiation showed increasingly marked degrees of inflammation (inflammatory cell infiltration, congestion, edema) and albuminous degeneration in the parenchymal cells. Those given the fractionated doses had more severe inflammatory and degenerative changes, culminating in fibrosis.

English et al.[102] studied the long term reaction of dog salivary glands to irradiation. Six mongrel dogs 13 to 17 kg in weight received 1000, 1500, and 1750 R of X-irradiation applied locally to the head. One was given the low dose, one the intermediate dose, and four the high dose. They were sacrificed between 5 and 16 months postirradiation. At necropsy, the salivary glands were promptly removed and those from one side were frozen in dry ice. Those from the other side were sliced and fixed in 10% formalin, Bouin's solution, Zenker's solution, Carnoy's solution, and Regaud's solution. Fixed tissues were dehydrated, embedded in paraffin, sectioned, and stained with hematoxylin and eosin, PAS, mucicarmine, Mallory's aniline blue collagen stain, Masson's trichrome, and Regaud's mitochondrial stain. Tissues from seven nonirradiated mongrel dogs served as controls.

Grossly, the irradiated animals manifested mild gingivitis to sloughing of the gingiva and buccal mucosa in 7 to 10 days. Erythema and ulceration of the exposed skin surfaces occurred within 1 to 2 weeks. Epilation began at 18 days with substantial hair loss by 33 days. There was marked reduction in size of salivary glands which made them difficult to locate. Microscopically, the parotid glands in the irradiated dogs showed a great reduction in parenchymal tissue. Lobes were diminished in size, and acini were shrunken and randomly missing. They were replaced by concentric rings of flattened dark blue nuclei with numerous free cells, or connective tissue with scattered cell remnants. The submaxillary glands contained few mucous acini. The rest of the cells were distributed either in nests or as free cells that bore no resemblance to salivary glands either in detail or arrangement. Nests of cells were little more than rings of irregular darkly basophilic staining bodies with small dense nuclei. Isolated cells and cell fragments were scattered throughout. Except for a local increase in connective tissue and some reduction in parenchymal tissue, the mixed sublingual gland acini appeared normal in some animals. Some showed a moderate amount of damage to the mucinous acini. In still others, destruction was so complete that most lobes contained only a few residual acini. The original pattern was entirely lost leaving masses of cells, nuclei, and cell fragments separated by broad bands of connective tissue. Whereas 1000 R caused minimal or recoverable changes, 1750 R produced atrophy, fibrosis, and marked architectural alterations in the dog salivary glands.

Mayo et al.[103] studied the effects of total body irradiation on the oral tissues of the Syrian hamster. In the first experiment, 110 (55 male, 55 female) 6- to 8-week-old hamsters weighing from 60 to 90 g were irradiated in groups of 10 with doses ranging from 400 to 2000 R. The last group served as nonirradiated controls. All were fed a pelleted diet, and each survivor was killed after 30 days. In the second experiment, 48 (24 male, 24 female) hamsters were divided into four groups. Groups I and II were exposed to 800 R total body irradiation. Group I was fed the pelleted diet; Group II received the same diet in a soft pappy form. Groups III and IV ate the same food regimens as Groups I and II, but were not irradiated. All were killed when either agonal, had oral lesions, or reached 30 days postirradiation.

Animals given doses of 700 R or higher underwent rapid deterioration with severe diarrhea, dehydration, and weight loss. Facial edema, suppurating blepharitis, and hemorrhagic exudate from the nose were common. Oral lesions were found in 30% given 700 R, in 90% subjected to 800 R, and in 100% exposed to 900 R or higher. The greater the dose, the sooner the appearance of the lesions. None became manifest before the 3rd day postirradiation. Lesions were yellow-gray and necrotic in appearance and were located in the lower lingual gingiva, with extensions to the floor of the mouth and around the incisors. In the upper mouth, they were located in the palatal gingiva of the incisors and molars. Each lesion was surrounded by an erythematous halo. Increase in size was associated with marked edema, hemorrhage, and suppuration. Often they formed fairly large abcesses. At the high doses, the lesions were more severe and involved the tongue. In animals surviving the 30-day experimental period, the oral lesions healed, leaving marked gingival recession, retractile scars, and obliteration of the mucobuccal fold. Microscopically, the lesions consisted of areas of necrosis confined primarily to the periodontal region that progressed to the oral mucosa, soft tissues, and alveolar bone margin. Irradiated animals given the soft pappy diet had more severe lesions than those ingesting the same diet in pellet form.

The effects of high doses of cobalt[60] irradiation on rhesus monkey mandible was examined with light microscopy by Rohrer et al.[104] Eight *Macaca mullata* monkeys, ranging in weight from 5.8 to 13.8 kg and in age from 1 to 10 years, were divided into control and test groups. Six were irradiated with cobalt[60] in fractionated doses of 450 R per day split bilaterally over a 12-day period for a total of 4500 R. The irradiation was similar in amount and method of delivery to that given human patients with cancer. The observation period ranged from 1 week to 6 months postirradiation.

Clinical observations paralleled the signs and symptoms in human patients given radiotherapy to the head and neck, notably, severe mucositis, soft tissue necrosis, alopecia, weight loss, fetor oris, sialorrhea followed by xerostomia, thick ropy saliva, anorexia, and cachexia. By 15 days postirradiation, the mucositis had almost completely disappeared and appetite returned. Microscopically, there was a loss of osteocytes from bone lacunae in the direct path of the beam in the outer lamellar and Haversian bone, but not in the cancellous bone. Other irradiation-related changes included loss of cellularity, vascularity and adherence of the periosteum; fibrosis; loss of principal fiber arrangement and decrease of vessels in the interstitial spaces in the periodontal ligament; fibrosis, loss of hematopoesis, proliferation of new bone and obliterative endarteritis in the bone marrow; and narrowing or obliteration of blood vessels and plugging of canals with osteoid in the Haversian systems. None of the monkeys developed any clinical signs of osteoradionecrosis during the 6-month postirradiation period.

Meyer et al.[105] compared the effects of 200 kV radiation and of cobalt[60] radiation on the jaws and dental structures of white rats. In their study, 70 male rats 3 months of age, weighing approximately 200 g, were used; 10 served as controls. The other 60 were divided into six groups of 10. Groups I, II, and III received 200 kV radiation at doses of 1020, 1530, and 2040 R; Groups IV, V, and VI received identical amounts of cobalt[60] irradiation. All radiation was given as a single bolus, and the dose was calculated as delivered to the right mandible. Surviving animals were killed after 2 and 4 weeks and the mandibles and maxillas examined histologically.

Animals subjected to a single dose of 200 kV radiation had damage to the odontoblastic layer and reduction in number of osteoblasts in the periodontal bone. There were severe alterations in the pulp and periodontal tissues. The periodontal membrane had a notable decrease in vascularity, with only occasional small sclerotic vessels in evidence. The fibroblast content of the membrane was reduced, and the cells that were present were small, rounded, and more deeply stained than normal. Cobalt[60]-irradiated animals had much less damage to the bone and periodontal tissues at comparable doses.

The histopathologic reaction of the cartilaginous growth center of the mandibular condyle to X-irradiation was elicited in the rat by Furstman.[106] In this study, 90 white male Holtzman rats averaging 45.7 g in weight were separated into three groups of 30 each. Group I was exposed to 600 R total body irradiation. Group II contained pairfed nonirradiated controls, and Group III nonirradiated *ad libitum* fed animals. All were given Purina® rat meal. Five animals from each group were killed at intervals between 1 and 28 days after irradiation of Group I. Animals were necropsied and the jaws embedded in nitrocellulose after fixing. Sections were stained with Mallory's connective tissue stain, PAS, and Alcian blue-PAS.

Total body irradiation seriously damaged the growing cartilage of the rat mandibular condyle. There was a high degree of osteoclastic activity within 24 hr after irradiation, together with an increase in osteoblasts, decrease in size and number of primary trabeculae and decided loss of hematopoetic elements in the bone marrow. Animals killed 3 and 5 days after irradiation displayed further decreases in thickness of the cartilaginous layers of the condyle and intensification of the changes in the 1-day postirradiation animals. Earliest signs of repair were noted 7 days after irradiation when proliferation of blood vessels into the diaphyseal part of the metaphysis was accompanied by reappearance of hematopoetic elements in the marrow spaces, decrease in fibrosis, increase in osteoblasts, and decrease in osteoclasts. Repair was well organized by 14 days postirradiation, as evidenced by a notable increase in osteoclastic activity in the region of primary trabeculae at the edge of the calcifying matrix and the almost complete recovery of the hematopoetic marrow elements. Repair was complete by 28 days postirradiation. Inanition did not affect the changes in the irradiated animals since they were not found in Group II animals.

Burstone[107] recorded the response of the mandibular joint of the mouse to varying doses of X-irradiation. In this study, 38 Swiss and C57 black mice, 2 to 7 days of age, were irradiated in the area of the mandibular condyle. Litter mates and litter mate controls were used when feasible. Animals received dosages of 1500, 3000, or 5000 R. Following irradiation, 24 surviving mice were sacrificed at intervals between 5 and 62 days. Litter mate controls were killed at the same time. Heads of mice under 3 weeks of age were fixed and decalcified in Bouin's solution with 10% acetic acid. Heads of older mice were split longitudinally and fixed in 10% formalin. All tissues were decalcified, embedded in paraffin, sectioned in a frontal plane, and stained with hematoxylin and eosin.

X-ray irradiation of the mandibular joint caused a marked inhibition in the ossification process. Irradiation with 1500 to 5000 R damaged the intermediate and hypertrophic zones of the condylar cartilage, with subsequent marrow aplasia and fibrosis. The cranial portion of the joint and the interarticular disc were relatively radioresistant. There was some restoration of growth potential approximately 6 weeks following irradiation with 5000 R, evidenced by marked increase in the cellular activity of resting and intermediate zones.

The effects of experimental irradiation on tooth development have been studied in salamanders, rodents, and swine. Brunst et al.[108] irradiated 5 axolotls (Mexican salamanders — *Siredon mexicanus*) with 2000 R 20 days after hatching, 5 with 4000 R 22 days after hatching, and 22 with 3000 R 60 days after hatching. Only the lower jaw was irradiated in each instance. The upper jaw was protected by lead shielding.

Enamel epithelium in the young axolotls disappeared after irradiation, and mitotic figures were completely suppressed in the surrounding tissues. Only small remnants of teeth were found 34 days after irradiation of the 20-day-old axolotls and 43 days after irradiation of the 60-day-old axolotls. All tooth remnants had disappeared 15 days later. There was no formation of new teeth following disappearance of the irradiated teeth by 1-year postirradiation. Both the mechanism of tooth formation and of tooth maintenance were permanently destroyed by X-irradiation in the doses used.

Burstone[109] described the impact of X-ray irradiation on the teeth and supporting structures of the mouse. In this study, 44 Swiss and C57 black mice aged 2 to 28 days were irradiated in the region of the molar teeth. Animals were exposed to 1500, 3000, or 5000 R and fed a Purina® lab chow. Litter mate controls were used whenever possible. The 32 mice that survived irradiation were sacrificed between 5 and 37 days. Controls were killed at the same time. Specimen preparation and staining were the same as described previously.[107]

Erythema and subsequent epilation of the irradiated skin occurred in 4 to 6 days after irradiation. At 3000 and 5000 R there was necrosis of the skin and underlying tissues. Administration of 1500 R to mice under 4 days of age caused hemiatrophy of the mandible and retardation in rate of eruption of the incisor tooth on the irradiated side. Radiographically, root development was absent on the side receiving the greatest irradiation. Animals irradiated with 5000 R over a small area at the apex of the incisor exhibited developmental anomalies expressed as spiked, malformed, or hypoplastic incisors and failure of incisors to erupt. Doses of 3000 to 5000 R produced a generalized retardation of tooth development in the experimental animals, as compared to the litter mate controls.

The degree of radiation damage to the teeth was dependent on the age of the mouse and stage of histiogenesis of the tooth. At 5000 R, the ameloblasts appeared as low cuboidal or squamous cells with a cessation of histodifferentiation of new ameloblasts. Doses of 1500 to 3000 R produced less marked changes in the enamel. All doses severely damaged the odontoblastic layer. Surviving cells resembled osteoblasts and had formed an irregular, amorphous, acidophilic substance (osteodentin). All doses also caused retardation or cessation of root formation in the first and second molars, while development of coronal dentin was only slightly altered. Odontoblasts were more radiosensitive than ameloblasts. Development of the basal and alveolar portions of the jaws was retarded or completely stopped. Late postirradiation effects included atrophy and fibrosis of the pulp and ankylosis of tooth root to alveolar bone.

The histologic changes in the rat upper incisor and supporting tissues produced by single doses of X-irradiation have been described by Medak et al.[110] In this study, 59 adult white rats, 30 to 90 days of age, 100 to 200 g in weight, were subjected to a single dose of X-irradiation ranging from 1000 to 4000 R. The period of survival was 6 hr to 43 weeks. Immediately after sacrifice, the heads were fixed in 5% formalin, radiographed, the upper incisors and surrounding tissues dissected out, blocked in paraffin or celloidin, sectioned, and stained with hematoxylin and eosin, silver impregnation, and Mallory's connective tissue stain.

At doses of 1000 to 2000 R, the changes were localized to the labial part of the odontogenic zone consisting of transient edema and permanent injury to the zonal odontoblasts that caused dentinal hypoplasia and produced dentinal niches. After degeneration of the odontoblasts, osteodentin was formed in the adjacent area of the pulp. Slight inhibition of pulpal growth led to waviness of the dentinoenamel junction. Slight injury to the ameloblasts caused shallow enamel hypoplasia. At doses of 3000 to 4000 R, destructive changes were followed by regeneration. Destruction was manifested by severe pulpal and periodontal edema, which led to formation of cystic cavities that destroyed the odontoblasts and prevented formation of new ameloblasts. Great masses of osteodentin formed in the pulp. With stoppage of pulpal growth and eruption, there was progressive maturation of the entire enamel matrix and reversion of the enamel organ into reduced epithelium. Regeneration was dependent on the amount of enamel destruction caused by the expanding cyst cavities. Only if viable remnants of odontogenic epithelium persisted to the time of organization of the cysts did epithelial proliferation initiate formation of a new incisor.

Subsequently, Medak et al.[111] examined the effect of single doses of X-irradiation on the eruption rate of the rat upper incisor. In this examination, 84 albino rats ranging in age from 30 to 90 days, and in weight from 80 to 200 g, were studied. Groups of eight animals were exposed to 500, 1000, 2000, or 2500 R. Twelve received 3000 R; 28, 4000 R; and 12 served as controls. The animals were anesthetized with Nembutal® and shielded with lead plates that left only the interior part of the head exposed. The central rays were projected toward the basal end of the upper incisor, and the full dose was given at one exposure. Rate of eruption of the upper incisor was measured in each animal. Animals given 500 and 1000 R showed no retardation in eruption. At 2000 R, eruption rate decreased during the first 4 weeks and then gradually increased to that of the animals given the lower doses. At 2500 R, slight stimulation in rate was followed by deceleration and cessation. At 3000 R and above, eruption stopped completely within 3 weeks.

The long term effects of a single dose of ionizing radiation on the rat incisor were reported by English et al.[112] Three litter mate groups of white rats were exposed to a single dose of localized head X-irradiation at 21 days of age. The maximum dose was 1500 R. Striking changes were found in the incisors of the irradiated animals sacrificed 100 days after treatment. The incisor teeth had separated into two segments. In the maxilla the first segment was frequently lost, leaving a stumplike tooth; in the mandible the second segment frequently grew lateral to the first producing an "extra" incisor tooth. Histologically, there was great damage to the tooth forming elements active at the time of irradiation, culminating in arrest of tooth formation.

Medak et al.[113] investigated the effect of X-irradiation on the incisors of the Syrian hamster and found that with the exception of the odontogenic epithelium, the reaction in the hamster was essentially the same as in white rats exposed to similar doses of irradiation. In the hamster, odontogenic epithelium that continued to grow after treatment was reduced to a thin strand of cuboidal or squamous cells. Local proliferation of the distorted strands of displaced epithelium resulted in the formation of several knoblike areas. The earliest stage of cellular proliferation produced an enlarged band of tissue without keratinization, but as proliferation continued, keratin was deposited. Later, an enlarged cyst filled with keratin was formed. In contrast to the rat, the odontogenic epithelium of the hamster did not resume normal function during the period of observation.

Kaplan and Bruce[114] investigated the influence of irradiation on the forming teeth of preweanling Syrian hamsters. Jaws of five groups of 10-day-old hamsters were exposed to a single dose of either 250, 500, 1000, 2000 or 4000 R X-irradiation. Each group contained six animals that were compared with six untreated litter mate controls. All animals were killed at 100 days of age and had roentgenograms taken. The magnitude of untoward effects was directly related to the dosage. At 250 R, there was little effect on tooth development; at 500 R, dwarfing of the molar teeth and incomplete root formation were evident; at 1000 R and above, irradiation damage was manifested by anodontia and dwarfed and rootless teeth.

English and Tullis[115] examined the developing teeth of 41 young swine grossly and histologically for pathologic changes resulting from exposure to ionizing radiation. The swine were selected from 20 animals exposed to 2000 kV total body irradiation in doses of 250 to 8000 R. A characteristic type of enamel hypoplasia was found in the developing molar teeth of animals given more than 400 R. Ameloblasts actively producing enamel matrix appeared to be more sensitive to irradiation than other parts of the teeth.

Cowgiel[116] followed the eruption of irradiation-produced rootless teeth in monkeys. In this study, 18 rhesus monkeys were subjected to fractionated external irradiation of the mandible and maxilla with either 4500, 5500, or 7500 R. The animals ranged in age from 14 to 28 months, and each had a complete deciduous dentition. Preirradiation radiographs revealed that the permanent teeth were in various states of formation. They were irradiated 5 days a week on alternating sides of the face for 5.3 to 12.5 weeks and lived up to 3.25 years after treatment. The monkeys were killed with an overdose of Nembutal® and the tissues fixed in 10% neutral formalin, decalcified, embedded in celloidin, sectioned, and stained with hematoxylin-eosin-Azure II. Tooth formation was followed in the postirradiation period by lateral jaw radiographs taken at regular intervals.

At all dose levels, irradiation completely inhibited further crown and root development and destroyed the anlages of the third molars. Calcification of the crown matrix was slowed, but continued to calcify after irradiation. Maxillary permanent teeth usually had some short thin root formation, but mandibular teeth were completely rootless. Pulp chambers of all teeth were reduced in size and the apex closed at the level of maximal tooth development at irradiation. Layers of dense bone formed under the erupting crowns. Most crowns erupted, while a few did not because of ankylosis in high-dose animals. Roots of the deciduous molars resorbed in the usual manner. Eruption rates of the partially formed bicuspid teeth were somewhat retarded.

All partially formed teeth eventually erupted into full occlusion. A few partly formed crowns were attached only to the gingiva and were in functional occlusion for 18 months without any evidence of looseness. Without root formation, dental follicles did not differentiate into periodontal ligament. They maintained attachment to the cemental surface of the crown, regardless of shape or size. Rootless teeth were held in place by connective tissue fibers from the follicular sac and were in functional occlusion. Eruption of the teeth appeared to be dependent on the growth of the follicular sac and/or alveolar bone.

Pratt et al.[117] tested the efficacy of radioprotection of the rat parotid gland by S-2(3-aminopropylamino) ethyl phosphothioate (WR-2721). This compound has been shown to be more protective for normal tissue than for tumor tissue against X-ray damage. Female Sprague-Dawley albino rats, 180 to 250 g in weight, were anesthetized with pentobarbital and given a single dose of 3200 R X-irradiation to the right side of the head from the midline laterally and from ear to clavicle while the other side was shielded; 15 to 20 minutes before irradiation the rats were injected i.p. with either saline or WR-2721 at a level of 400 mg/kg — 60 days after irradiation the rats were fasted overnight, reanesthetized, and sacrificed by perfusion-fixation through an aortic cannula. After perfusion, the glands were excised, immersion fixed for 5 hr, and washed overnight in several changes of sodium cacodylate buffer containing sucrose. They were then postfixed in osmium tetroxide, dehydrated in alcohol, and embedded in resin. Sections were double-stained with uranyl acetate and lead citrate and examined by electron microscopy.

Nonprotected glands suffered a drastic reduction in amount of acinar tissue while the ducts and blood vessels showed only minor morphologic changes. Evidence of acinar loss was provided by the presence of excessive and redundant basal laminae around surviving acini, and by large concentrations of empty basal laminal profiles in the spaces between surviving structures. Blood vessels present at 60 days postirradiation showed no evidence of damage. There were, however, additional layers of basal lamina around occasional capillaries which was interpreted as evidence of previous endothelial repair. The WR-2721 protected glands showed similar signs of damage, but to a much lesser degree. The greatest protection was provided to the radiosensitive acinar cells.

In reviewing what is and what, hopefully, will be the only such "naturally occurring experiment" in human history, DeCoursey[118] detailed the effects of atomic bomb radiation on the mouth structures of Japanese victims of Hiroshima and Nagasaki. Lesions of the oral regions became of utmost clinical importance between the 3rd and 6th week after the explosions. Grayish bullae appeared in the posterior pharynx followed by petechial or ecchymotic mucosal hemorrhages with superimposed ulcerations. Ulcerations similar to those of agranulocytic angina developed over the tonsillar areas and spread from one lymphoid island to another, ultimately involving the tongue. The gingivae were hemorrhagic and ulcerated with tissue destruction sometimes spreading to the buccal mucosa, involving the entire cheek and producing the clinical picture of noma. Almost all patients with such lesions died.

EXPERIMENTAL OSTEORADIONECROSIS OF THE JAWS

Cowgiel[119] studied the quantitative relationships between radiation dose, bone sensitivity, and pathogenesis of osteoradionecrosis in the jaws of 20 rhesus monkeys 15 to 18 months of age with a mixed dentition. Three groups of six each were given either 4500, 5500, or 7500 R, and two served as controls. Three animals from each group were subjected to a second course of irradiation after an interval of 2 to 3 months in amounts of 3500 to 4000 R. At the completion of the experiment, the monkeys were killed and the jaws fixed in 10% formalin, decalcified, blocked, embedded in celloidin, sectioned, and stained with hematoxylin and eosin-Azure II.

The time of onset of osteoradionecrosis was directly related to the radiation dose. In the 11,000 R group, necrosis was present 1 day postirradiation; in the 9500 R group, 1 week postirradiation; in the 8500 R group, 2 weeks postirradiation, and in the 7500 R group, 6 to 14 weeks postirradiation. Although necrosis did not occur in the 5500 R group, one animal in the 4500 R group developed the disease 19 months after irradiation. In the 9500 to 11,000 R dose range, necrosis followed a very rapid course, was usually bilateral, involved the tongue and maxilla, and perforated the cheek. At 7500 and 8500 R, necrosis developed slowly, was usually unilateral and involved the maxilla, tongue, and cheek in half the animals. In all instances, the necrosis followed a similar pattern beginning in the interdental papilla between the lower second deciduous molar and the first permanent molar on one side. It extended to the cervical gingiva, attached gingiva, and buccal and lingual mucosa. Secondary infection accelerated the cellulitis and caused early sequestration of bone. Bone changes always followed gingival breakdown.

Histologically, the initial events were lymphocytic infiltration and dilatation of the blood vessels in the corium of the interdental papilla. The epithelial attachment proliferated downward along the root surface, became thin and infiltrated with lymphocytes before undergoing ulceration and necrosis. Inflammation and necrosis spread along the periodontal ligament involving contiguous structures by direct extension. Spread in the gingiva was by direct extension under intact epithelium. Regardless of route, the necrosis was always in advance of the gross lesion.

In bone, osteocytes adjoining the necrotic area became pyknotic while those two or more lacunar spaces away were usually normal appearing. Marrow spaces adjoining the necrotic area became filled with chronic inflammatory cells. Osteoclastic activity was prominent at the periphery of the lesion. Vascular changes were found mainly in the mandible where arterioles and, occasionally, medium- and large-sized arteries were involved. The intima and media were thickened by cell proliferation producing slight to complete occlusion. Bone destruction in the alveolar part of the mandible stimulated a periosteal reaction along the inferior border that resulted in new trabeculae being laid down on the cortical plates. The amount of new bone was proportional to the amount destroyed. Osteoradionecrosis resulted primarily from radiation damage to the osteocytes and was aggravated and prolonged by blood vessel changes.

Bond et al.[120] assessed the influence of regional oxygenation on experimental osteoradionecrosis. Six adult rhesus monkeys (4 males, 2 females) ranging in weight from 8.5 to 11.5 lb received identical amounts of radiation to the right and left mandible, followed by extraction of the lower first molars. A 50-m*l* intra-arterial infusion of 0.12% hydrogen peroxide was made through an indwelling catheter in the right common carotid artery. Each side was then exposed to a single dose of 4750 R delivered to the first molar area. The lower first molars were extracted from all monkeys between 61 and 63 days postirradiation. At surgery, buccal plate and interseptal bone were taken for histologic examination. The animals were killed 177 to 199 days postirradiation and the mandibular extraction sites removed and prepared for histologic examination.

At sacrifice, little tendency to healing was found in the extraction sites and varying degrees of osteonecrosis were evident clinically. Histologic examination of both sides revealed osteoradionecrosis in the compact bone of the cortical plate denoted by osteocyte death and degeneration, empty lacunae, absence of osteoblasts, presence of a few osteoclasts, and endarteritis and periarteritis in the Haversian systems. On the control side, the bone marrow showed fibrous degeneration and osteoid reduction. On the infused side, there was a loose stroma of adipose and areolar tissue containing myeloid elements and a proliferation of young cancellous bone within the subperiosteal layer external to the nonvital cortical plate. The expected bone marrow suppression following irradiation was significantly reversed by regional oxygenation. Hydrogen peroxide provided a stimulus for proliferation of cancellous bone subperiostally, endosteally, and within the pulp. Hydrogen peroxide also prevented the endarteritis and periarteritis associated with osteoradionecrosis.

ORAL REACTIONS TO EXPERIMENTAL TRAUMA

Shteyer and Howell[121] compared the response to chronic irritation of the tongue in germ-free and conventional rats. In the study, 32 germ-free Fischer albino rats and an equal number of conventional rats of the same strain were anesthetized with pentothal. The lower right first and second molars were separated with a modified towel clamp to permit encirclement of the first molar by a stainless steel wire. Free ends were twisted together on the lingual aspect to form a tight ligature. The twisted end was cut to a length of about 4 mm and directed into the lateroventral surface of the tongue, where it remained until the animal was killed. The rats were sacrificed at intervals between 1 and 28 days after ligation. The wounds were examined grossly before the tongues were removed, fixed in 10% formalin, and processed for preparation of hematoxylin and eosin-stained slides.

Tissue reactions during the first 2 days of irritation were histologically similar in both groups. After the 3rd day, the conventional rats had a denser inflammatory cell infiltrate at the wound edges. At 7 days, the germ-free animals evidenced granulation tissue proliferation and walling off of the wound, whereas in the conventional animals the inflammatory process continued to spread. By day 14, maturation of granulation tissue in the conventional rats approached that of the germ-free animals. At 28 days, the tongue wounds were comparable except for increased granulation tissue in the conventional rats. None of the wounds in either group underwent reepithelialization during the course of the study.

Bhaskar and Lilly[122] described the histiogenesis and natural history of traumatic granuloma of the tongue that in man lasts for a few days to a few weeks, ulcerates, and contains histiocytes, neutrophils, and eosinophils. In this study, 58 albino rats, 8 weeks of age, were divided into two experimental groups. Group I, 26 animals, was anesthetized with ether and the tongue subjected to a single crushing injury with specially designed forceps equipped with rubber tips to prevent laceration. The area of trauma was 5 mm wide and located in the middle third of the tongue. Pressure was controlled by a built-in stop on the handle. Group II, 26 rats, was treated like Group I plus repeat of the trauma to the same area in the same manner for six consecutive days. The animals were killed in groups of three between 0 and 21 days following injury. Tongues were excised, washed, formalin fixed, and processed for staining of sections with hematoxylin and eosin.

In the single-trauma animals, the sequence was hematoma, edema, degeneration of muscle fibers, formation of giant cells, proliferation of histiocytes, infiltration of neutrophils, differentiation and proliferation of fibroblasts, and formation of an irregular scar. In the multiple-trauma animals, lesions developed in two ill-defined phases. In phase one, there was an irregular area of histiocytic proliferation and neutrophil infiltration with degeneration of the muscle fibers. The lesion extended into the surrounding musculature presenting a pseudoinvasive appearance that resembled the human counterpart. In phase two, there was a progressive increase in the number of fibroblasts and collagen fibers, with a decrease in histiocytes and neutrophils that ended in scar formation.

Chierici et al.[123] developed a primate model to study alteration in facial morphogenesis in response to experimental deformation of the maxilla. Complete unilateral surgical clefts of the premaxilla and hard palate were produced in 30 rhesus monkeys. The clefts extended from the posterior border of the hard palate anteriorly through the alveolar ridge. They were about 3 mm wide posteriorly and somewhat wider anteriorly. Central and lateral incisors were removed on the operated side and the labial flap of the mucosa sutured closed. A compression appliance was wired to two teeth on each side of the upper arch to produce pressure on the bony edges of the defect and to induce a medialward deformation of the maxillary segment. All monkeys recovered rapidly from the procedure, remained healthy and continued to grow. At periodic intervals they were anesthetized to check the progress of maxillary collapse.

The monkeys developed an obvious asymmetry of the upper jaw reflected in the soft tissues of the face. There was a deviation of the nasal septum with an altered position of the ala of the nose on the affected side. The maxilla on the operated side rotated around an axis located behind the tuberosity resulting in a decided crossbite in the collapse segment and producing significant changes in the position and occlusion of the dentition. They did not extend to the zygoma indicating that this bone is not part of the buttress system influenced by forces acting through the dentition.

The effects of laser beams directed on the teeth and oral mucosa of Syrian hamsters was investigated by Taylor et al.[124] Twelve animals aged 2 to 3 months were anesthetized with Nembutal®. Six received 35 J from a laser beam focused on the buccal surface of the mandibular left or right incisor and on the lateral borders of the tongue on the same side. Six were exposed in the identical manner to 55 J. Half the animals were killed 3 days after exposure; half, 7 days. The oral structures were processed for histologic scrutiny of hematoxylin and eosin, Mallory's connective tissue, and PAS-stained sections.

Cavitation was produced in the enamel and dentin of the mandibular incisors. Adjacent to the cavitation, normal appearing enamel was replaced by a chalky white, amorphous substance in the 35-J animals and by a smooth, glassy substance in the 55-J animals. Initially, there was a small reddish brown lesion at the site of the tongue exposed to 35 J. After 3 days, a sharply demarcated large ulcer was manifest. Adjacent mucosa was erythematous and edematous. In hamsters given 55 J, the ulcerated area on the tongue was deeper, with greater reddening and swelling of the adjacent mucosa at 3 days than in those given the lower dose. Lingual papillae were flattened, and the entire tongue was enlarged and swollen. At 7 days, the tongue was still edematous, and healing was less advanced than in animals given the lower exposure. Microscopically, the tongue ulcers had the typical pattern of nonspecific ulceration, with acute purulent inflammation. Evidence of granulation tissue formation and epithelialization was present at 7 days. Tongues exposed to 55 J showed disruption of muscle bundles and destruction of striated muscle tissue. Epithelium was thin and papillae were atrophied. The dental pulps of the irradiated incisors had severe degenerative changes. Less severe pulpal degeneration was evident in the molar teeth located at some distance from the laser focal point.

The experimental production of dental ankylosis in dogs by means of trauma was attempted by Parker et al.[125] Conditions producing ankylosis in man often involve tissue injury, immobility, and lack of function. In the study, 16 dogs were anesthetized and the teeth examined radiographically. Splints were fabricated and cemented to immobilize selected teeth. Tissues investing the teeth were reflected and adjacent bone mechanically injured. The teeth were then ground out of occlusion. Although subsequent radiographic studies indicated that many of the treated teeth had become ankylosed, true ankylosis or continuity of bone with tooth substance was found in only one animal on histologic examination.

Butcher and Klingsberg[126] examined the influence of age, and of gonadectomy on the healing of experimentally induced wounds in the palatal mucosa of the rat. Almost 200 Long-Evans rats of both sexes ranging in age from 22 to 830 days were studied. Palatal wounds were made by two methods in pentobarbital-anesthetized animals. In the first, a trephine 1.5 mm in diameter was used to punch out circular sections involving only the superficial mucosa, and deep wounds, including the periosteum. The second involved surgical removal of an area of mucoperiosteum 2 mm long and 1 mm wide from the palate medial to the first and second molars. Biopsy specimens of the wound sites and adjacent tissues were taken from different rats on successive days following injury, fixed and sectioned for microscopic examination. Gonadectomies were performed in animals of different ages. After several days or months, palatal injuries were made in these animals in the same way as in the intact animals.

Superficial wounds in the oral mucosa followed the same healing sequence as wounds in skin where epithelial cells move over the connective tissue. Deep wounds were filled in with granulation tissue from the sides that carried epithelium. The superficial portion of the exposed bone often degenerated, producing a sequestrum that was resorbed, undermined, and expelled by growth of granulation tissue. Palatal wounds healed more rapidly in 50-day-old rats than in 200-day-old rats. In the aged animals, the wounds healed very slowly and left many defects. Healing appeared to progress more rapidly in ovariectomized animals than in those subjected to orchiectomy.

EXPERIMENTALLY INDUCED ENAMEL HYPOPLASIA

Enamel hypoplasia has been produced in experimental animals by hormone depletion (parathyroidectomy, thiouracil administration), vitamin deficiencies (vitamin D, vitamin C, vitamin B_6), mineral deficiencies (magnesium, phosphorus), and by fluoride excess.[127] As demonstrated by Kreshover,[128] enamel hypoplasia can also be caused in developing teeth of rodents by infections that damage the ameloblasts and interfere with amelogenesis. In this study, 70 mice, 10 to 50 days old, were injected i.p. with either bovine or human type tubercle bacillus. Amount of inoculum ranged from 0.05 to 0.7 mg/100 g body weight. Serving as controls were 22 noninfected mice. The animals were sacrificed between 186 and 204 days after inoculation. The mandibles of all 92 mice were decalcified, celloidin- and paraffin-embedded, serially sectioned, and stained with hemotoxylin and eosin. In a corollary study, 14 guinea pigs were inoculated either i.p. or in the groin with human type tubercle bacillus. Inoculation doses were from 0.05 to 0.1 mg/100 g body weight. All were fed a normal diet. Time of death from either natural causes or by sacrifice was between 8 and 160 days. At necropsy, the mandibles were removed, halved, and fixed in 10% formalin. After double-embedding (celloidin and paraffin), serial sections were prepared and stained with hematoxylin and eosin.

Dental changes in the experimental animals were strikingly similar to those described in human material. Microscopically, there were abnormal formation or secretion of enamel-like substances subjacent to and within the cell cytoplasm, vacuolization of ameloblasts, complete cystic destruction of ameloblasts, and cessation of enamel formation, resulting in notchlike defects. Lesions were found in both incisors and molars. Dentinal changes were represented by wide prominent incremental lines reflecting long periods of arrested development. In many instances, the odontoblastic layer in the root portions of the teeth was absent. Subjacent pulp contained large irregular masses of pink-staining osteoid tissue in which tubules were frequently present. Many of the odontoblasts showed marked cystic changes.

EXPERIMENTAL STUDIES ON ORAL SENSATION

Taste is unique among the chemical senses in being associated with discrete receptor organs of nonneural derivation, the taste buds. Classically, taste has been subdivided into four modalities, each localized to a specific portion of the tongue. Sweet reception occurs in the fungiform papillae at the tip of the dorsum; salt, in the fungiform papillae on the lateral aspects of the anterior two thirds of the dorsum; bitter, in the circumvallate papillae; sour, in the foliate papillae on the lateral margins of the posterior one third of the dorsum.

Farbman[129] studied differentiation of oral epithelium and taste bud development in the fetal rat tongue. The epithelial cells that participate in the formation of taste buds are made up of dark and light elongated spindle-shaped cells that taper into narrow distal processes and end in a small supporting cell. The dark cells contain prominent secretory granules and are the secretory supporting cells; the light cells contain many mitochondria and large amounts of agranular endoplasmic reticulum and are the receptor elements. Taste bud cell population undergoes constant renewal. Taste buds depend upon an intact nerve supply for existence. When the nerve is interrupted, taste buds degenerate; when the nerve regenerates, taste buds reappear. Epithelial innervation is present in the rat tongue at 15 to 16 days *in utero*. The nerve invades as a sizable bundle of nerve processes into a small focus of epithelium and becomes contiguous with three to four basal cells that develop into taste bud cells. They have to be touched by the nerve to undergo differentiation. Differentiation into the two types of cells comprising the taste bud occurs by a process that is simultaneous with, but independent of, other oral epithelial cells.

Trefz[130] investigated the hypothesis that intracellular enzymes contribute in the transduction of tastes to electrical impulses by the taste cells. A histochemical survey was made in rhesus monkeys of the activity of 14 enzymes in taste buds associated with sweet, salt, sour, and bitter reception. All monkeys were killed by vascular perfusion of isoosmotic saline while under anesthesia. Tongues were incised posterior to the circumvallate papillae, and immediately thereafter regions containing taste buds were removed from well within areas classically assigned to each modality. Tissues were frozen in liquid nitrogen and sectioned on a cryostat; 14 different enzyme systems were analyzed histochemically in the tissue sections, as follows: for Krebs cycle activity — succinic dehydrogenase; for electron transport chain activity — choline oxidase, cytochrome oxidase, α-glycerophosphate dehydrogenase, β-hydroxybutyric dehydrogenase, and D-amino acid oxidase; for membrane-bound hydrolytic activity — nonspecific esterase; for membrane-bound pentose-phosphate shunt activity — NAD diaphorase and NADP diaphorase; for soluble pentose-phosphate shunt activity — glucose-6-phosphate dehydrogenase; for mitochondrial activity — adenosine triphosphatase; for Golgi complex activity — nucleotide diphosphatase; for active transport activity — alkaline phosphatase; for lysosomal activity — acid phosphatase.

Each of the enzymes studied was present in the tastebud cells, but response in the surrounding epithelium from which the taste cells were derived was variable. The enzymatic activity in the taste buds is shown in the following table:

Enzymatic Activity in Taste Buds

Enzyme	Sweet	Salt	Sour	Bitter
Succinic dehydrogenase	—	—	+ +	+ + + +
Choline oxidase	+ + +	+	+ + + +	+ + +
Cytochrome oxidase	+	+	+	+ +
Nonspecific esterase	+ +	—	+	+ + +
NAD diaphorase	+ +	+ + +	+ + + +	+ + + +
NADP diaphorase	+	+	+ + +	+ + + +
α-Glycerophosphate dehydrogenase	+	+	+ + +	+ + + +
β-Hydroxybutyric dehydrogenase	—	—	—	+ + +
D-amino acid oxidase	+ + +	+ + +	+ + +	+ + +
Alkaline phosphatase	+ + + +	+ + + +	+	—
Glucose-6-phosphate dehydrogenase	+ +	—	+ +	+
Acid phosphatase	+ +	+ +	+ + + +	+ + + +
Adenosine triphosphatase	+ +	+ +	+	+ + +
Nucleotide diphosphatase	+ +	+ +	+ +	+ + +

The hypothesis that specificity and transduction of the taste impulse is due to specific enzyme involvement was confirmed. Functional association was noted between sweet and salt enzymatic activity and between sour and bitter enzymatic activity. A bitter specific enzyme β-hydroxybutyric acid dehydrogenase was identified.

There are two important theories for activation of afferent nerves in teeth that carry impulses for pain recognition. The odontoblast and its process in the dentinal tissue may function as a neurosensory terminal for afferent nerve fibers in the pulp, or the odontoblast and its process in the dentinal tubule may provide a means for mechanically coupling the disturbance in dentin with afferent nerves in pulp. Although anatomical studies have delineated the pathway for pulpal nerves that course toward the CNS, identification of the "painful" afferent impulses has been obscured by the large number of nerve fibers mediating touch, temperature, and pressure that arise in the periodontal membrane and run in the same nerve bundles as those from pulp.

The nerves in the brain stem that are activated by noxious stimuli applied to teeth have been localized and identified by Miller et al.[131] Adult male and female cats were anesthetized with pentobarbital. Heads were fixed in position with ear pins and a dorsal midline incision was extended caudally from the frontoparietal symphysis of the skull to a midcervical level. Cervical muscles were detached from the occipital bone and from the neural arches of the first two cervical vertebrae. The dura mater covering the posterior surface of the medulla oblongata was exposed with rongeurs. A reference electrode was inserted into the cut surface of one of the cervical muscles and impalement made with the recording microelectrode into the medulla at a point directly lateral from the obex into the dorsal sulcus. This point was chosen because the electrode could be inserted into the descending tract of the trigeminal nerve. After each recording session, the cats were killed and the location of the recording electrode determined by histologic sections cut frozen and stained with thionine. Ice, heat, vacuum, pressure, and electric stimuli were applied with appropriate mechanical and electrical devices to cavity preparations extending into the dentin and the neural responses in the medulla observed or recorded with an oscilliscope.

Pain and temperature from the face and oral cavity were conveyed to the somatic sensory regions of the cortex by the descending tract of the trigeminal nerve. Electrophysiologic studies of nerve activity in this pathway offered an opportunity to study the neurophysiologic response to pain-eliciting stimuli applied to exposed dentin. Heat, electric shock, potassium iodide solution, and negative pressure increased the firing frequency of a majority of the neurons in the descending tract of the trigeminal nerve. Cold and positive pressure applied to the exposed dentin reduced the activity in many neurons with increases in a few.

EXPERIMENTALLY PRODUCED ORAL DISEASE WITH HUMAN COUNTERPARTS

Spouge and Cutler[132] investigated induced oral mucosal hypersensitivity lesions in laboratory animals in view of the positive statistical relationship between recurrent oral ulcerations and hypersensitivity disease in man. The basic objective was to assess oral mucous membrane susceptibility to initial instigation of the hypersensitive state and vulnerability of the sensitized mucosa to subsequent lesion production. Rabbits were used for studies of oral mucous membrane susceptibility to the immediate type of reaction (localized ulceration of the Arthus type) because they develop a high titer of circulating antibodies. One group was sensitized with horse serum (1:10 dilution) antigen by s.c. injection and another by submucosal injection. As soon as any positive sign of hypersensitivity (edema or erythema) appeared at the injection site, a provocative dose of allergen was injected into the skin or oral mucosa. Guinea pigs were used to demonstrate delayed hypersensitivity lesions because the Arthus phenomenon which mars development of the delayed reaction does not develop easily in this species. One group was sensitized via the skin and another via the oral mucosa with horse serum allergen mixed with Freund's adjuvant. As soon as any sign of delayed hypersensitivity occurred, a provocative dose of antigen was injected into the skin and mucosa to determine if both sites had become sensitized.

The sensitizing antigen was equally potent when administered via the skin or oral mucosa, producing immediate and delayed type hypersensitivity in the respective animals. Both the skin and mucous membrane became sensitized simultaneously, regardless of original site of sensitization. Provocative injections of antigen produced delayed type lesions with equal facility in the skin and mucous membranes. An appreciably greater amount of antigen was required to produce lesions of the immediate type (clinical ulcerations) in the oral mucosa than in the skin.

The classic constellation of findings in the Pierre Robin syndrome is micrognathia, glossoptosis, and cleft palate associated with respiratory obstruction. Along with these may be other minor malformations such as syndactyly or major anomalies such as achondroplasia, congenital amputations, and congenital heart disease. Although the exact etiology of the Pierre Robin syndrome remains obscure, Cooke[133] has successfully reproduced the syndrome in chickens. Fertile white Leghorn eggs were incubated at 38°C. From 0.25 to 0.50 mg deoxyguanosine in a suspension of 0.5% carboxymethylcellulose was injected into the yolk sac on the 4th day of incubation by needle puncture into the blunt end of 40 eggs that were candled daily. From the 10th day on, dead embryos were inspected and gross abnormalities recorded. Embryos surviving the 18th day were also examined.

There were no defects or malformations in a control series of 40 uninjected eggs. Of the 40 injected embryos, 7 died before — and 19 after — the 10th day of incubation. Cleft palate and micrognathia with glossoptosis were found in 12. Defects in the maxilla and palate were present in the other seven. The tongue developed from the mandible and was ptotic in all micrognathic embryos. Embryos surviving 18 days of incubation were sacrificed and examined. All 14 had beak and palate abnormalities, notably, cleft palate, micrognathia, maxillary cleft, and combinations thereof. The time of injection of the purine deoxyguanosine corresponded very closely to that of mandibular development in man. The resultant malformations mimicked those of the Pierre Robin syndrome.

Enclavement of epithelium with lymphoid tissue and nonneoplastic proliferation have been considered etiologically significant factors in the pathogenesis of lymphoepithelial lesions like branchial cleft cysts and papillary cystadenoma lymphomatosum. In line with this hypothesis, Vickers and von der Muhl[134] determined the inducibility of lymphoepithelial lesions by autogenous epithelial tissue transplanted into the cervical lymph nodes of Syrian hamsters. In this study, 20 male Syrian golden hamsters about 6 weeks of age were divided into two groups of 10 each. They were anesthetized with i.p. pentobarbital and the salivary glands, lymph nodes, and adjacent structures exposed by blunt dissection.

Group I animals received an autogenous transplant of 1.0 × 0.5 mm of freshly excised submandibular salivary gland into a submandibular lymph node on the right side. The lymph node was pierced with a 27-gauge, ½-in. needle prior to implantation. Similar exposure and instrumentation were performed on the left submandibular lymph nodes without implantation of autogenous tissue. Group II animals received an autogenous transplant of a 1.0 × 0.5 mm freshly excised portion of the buccal pouch into a submandibular lymph node of the right side. As in Group I, a similar perforation was made into the substance of a submandibular gland on the left side without inserting epithelium. The wounds were closed and the animals were killed after 30 days.

Nodes and glands were removed, fixed in 10% formalin, processed, and stained with hematoxylin and eosin.

Surgically transplanted submandibular salivary gland tissue largely disappeared during the experimental period. The transplant was present in two of nine animals without any evidence of cyst formation. In contrast, the transplants of buccal pouch epithelium into the lymph nodes of nine animals developed into keratinizing stratified squamous epithelium-lined cysts in seven instances. This supports the epithelial enclavement within lymph node origin of the lymphoepithelial cyst.

Dental changes occur in children with jaundice, especially that caused by hemolytic disease. Discoloration of deciduous teeth has been reported in almost 10% — and enamel hypoplasia in 80 to 100% — of children with kernicterus. To elicit the effect of experimentally induced obstructive jaundice on the teeth, Hals and Nielsen[135] produced biliary obstruction in 12 hamsters and in 52 young female rats by ligation and resection of the main bile duct. The animals were killed at varying intervals after surgery. Nonoperated and sham-operated animals served as controls. Most of the experimental group developed hyperbilirubinemia that remained constant throughout the study period. Bilirubin levels ranged from 5 to 15 mg% with conjugated bilirubin predominating. Green pigment (biliverdin) was deposited in the jaw bones and kidneys of the rats. In some instances, bile pigments were detected in the predentin and preenamel. Incursion of pulp tissue into the dentin was a common finding in hyperbilirubinemic animals of both species. Formation of free cementicles in the periodontium of the molars was a feature peculiar to the hamsters.

Bilateral noninflammatory enlargement of the salivary glands is a feature of the chronic phase of Chagas disease in man. The changes are due to hypertrophy of the acini. Such patients frequently manifest amylasemia and an increased amylase content in parotid gland tissue. To determine the mechanism for the salivary gland changes, Aleves and Machado[136] studied submandibular gland involvement in the acute and chronic phases of experimentally produced *Trypanosoma cruzi* infection in rats.

Male and female Holtzman rats, 27 to 28 days of age, were inoculated i.p. with 0.15 ml blood containing 300,000 trypomastigotes of the Y strain of *T. cruzi*. Infection was confirmed 11 days after inoculation by the presence of living trypomastigotes in the blood of all inoculated rats. Six groups of infected and litter mate control rats were killed under ether anesthesia. Both submandibular glands were dissected out and weighed. Slices of the right submandibular gland were fixed in Helley's fluid, washed, embedded in paraffin, sectioned, and stained with hematoxylin and eosin and with Gomori's trichrome. Submandibular gland tissue components, i.e., acinic and striated duct cells, granular duct cells, intercalated ducts, lumen of intralobular ducts, and stroma were counted in ten microscopic fields randomly selected inside the gland lobules of Gomori-stained sections.

The submandibular glands of young rats killed at intervals after inoculation with the Y strain of *T. cruzi* showed accelerated acinar development and retarded duct system maturation, particularly in the developing granular ducts. Acinar hypertrophy was evident 18 days after inoculation (acute phase); impairment of duct development was found up to 60 days after infection. No histologic or quantitative changes were found in glands of animals killed 10 to 12 months after inoculation (chronic phase).

REFERENCES

1. Ziskin, D. E. and Blackberg, S. N., The effect of castration and hypophysectomy on the gingivae and oral mucous membranes of rhesus monkeys, *J. Dent. Res.*, 19, 381, 1940.
2. Dickler, E. H., Toto, P. D., and Gargiulo, A. W., Cell kinetics of the oral epithelium of adrenalectomized and hypophysectomized mice, *J. Periodontol.*, 40, 31, 1969.
3. Carbone, D. F., Sweeney, E. A., and Shaw, J. H., The effect of hypophysectomy in the weanling rat on size of the submandibular gland and on salivary proteins, *Arch. Oral Biol.*, 12, 721, 1967.
4. Kronman, J. M. and Chauncey, H. H., Hormonal influence on rat submandibular gland histochemistry, *J. Dent. Res.*, 43, 520, 1964.
5. Shklar, G. and Chauncey, H. H., Effects of hypophysectomy on the enzyme histochemistry of the rat submaxillary gland, *J. Dent. Res.*, 42, 71, 1963.
6. Bixler, D., Webster, R. C., and Muhler, J. C., The effect of testosterone, thyroxine and cortisone on the salivary glands of the hypophysectomized rat, *J. Dent. Res.*, 36, 566, 1957.
7. Clark, P. G., Shafer, W. G., and Muhler, J. C., Effect of hormones on structure and proteolytic activity of salivary glands, *J. Dent. Res.*, 36, 403, 1957.
8. Schour, I. and van Dyke, H. B., Changes in the teeth following hypophysectomy. I. Changes in the incisor of the white rat, *Am. J. Anat.*, 50, 397, 1932.
9. Ziskin, D. E., Applebaum, E., and Gorlin, R. J., The effect of hypophysectomy upon the permanent dentition of rhesus monkeys, *J. Dent. Res.*, 28, 48, 1949.
10. Ziskin, D. E. and Applebaum, E., Effects of thyroidectomy and thyroid stimulation on growing permanent dentition of rhesus monkeys, *J. Dent. Res.*, 20, 21, 1941.
11. English, J. A., Experimental effects of thiouracil and selenium on the teeth and jaws of dogs, *J. Dent. Res.*, 28, 172, 1949.
12. Goldman, H. M., Experimental hyperthyroidism in guinea pigs, *Am. J. Orthod. Oral Surg.*, 29, 665, 1943.
13. Woollam, D. H. M. and Millen, J. W., Influence of thyroxine on the incidence of hare lip in the "Strong A" line of mice, *Br. Med. J.*, 1, 1253, 1960.
14. Hoskins, W. E. and Asling, C. W., Influence of growth hormone and thyroxine on endochondral osteogenesis in the mandibular condyle and proximal tibial epiphysis, *J. Dent. Res.*, 56, 509, 1977.
15. Schour, I. and Rogoff, J. M., Changes in the rat incisor following bilateral adrenalectomy, *Am. J. Physiol.*, 115, 334, 1936.
16. Liu, F. T. Y. and Lin, H. S., Role of adrenocortical hormones in growth and development of salivary glands in immature rats, *J. Dent. Res.*, 48, 602, 1969.
17. Che-Kuo, H. and Johannessen, L. B., Skeletal changes in cortisone treated male rats, *J. Dent. Res.*, 49, 34, 1970.
18. Dreizen, S., Levy, B. M., and Bernick, S., Studies on the biology of the periodontium of marmosets. X. Cortisone induced periodontal and skeletal changes in adult cotton top marmosets, *J. Periodontol.*, 42, 217, 1971.
19. Selye, H., Effect of cortisone and somatotrophic hormone upon the development of a noma-like condition in the rat, *Oral Surg. Oral Med. Oral Pathol.*, 6, 557, 1953.
20. Schour, I., Changes in the incisor of the thirteen-lined ground squirrel (*Citellus tridecemlineatus*) following bilateral gonadectomy, *Anat. Rec.*, 65, 177, 1936.
21. Shafer, W. G. and Muhler, J. C., Effect of gonadectomy and sex hormones on the structure of the rat salivary glands, *J. Dent. Res.*, 32, 262, 1953.
22. Nutlay, A. G., Bhaskar, S. N., Jr., Weinmann, J. P., and Budy, A. M., The effect of estrogen on the gingiva and alveolar bone of molars in rats and mice, *J. Dent. Res.*, 33, 115, 1954.
23. Liu, F. T. Y. and Lin, H. S., Effect of progesterone on growth and development of submandibular glands in female rats, *J. Dent. Res.*, 48, 943, 1969.
24. Kreshover, S. J., Clough, O. W., and Bear, D. M., Prenatal influences in tooth development. I. Alloxan diabetes in rats, *J. Dent. Res.*, 32, 246, 1953.
25. Anapolle, S. E., Albright, J. T., and Craft, F. O., Microvascular lesions of gingival and cheek pouch tissue in the diabetic Chinese hamster, *J. Periodontol.*, 48, 341, 1977.
26. Golub, L. M., Schneir, M., and Ramamurthy, N. S., Enhanced collagenase activity in diabetic rat gingiva: in vitro and in vivo evidence, *J. Dent. Res.*, 57, 520, 1978.
27. Liu, F. T. Y. and Lin, H. S., Role of insulin in body growth and the growth of salivary and endocrine glands in rats, *J. Dent. Res.*, 48, 559, 1969.
28. Mangos, J. A., Benke, P. J., and McSherry, N., Salivary gland enlargement and functional changes during feeding of pancreatin to rats, *Pediatr. Res.*, 3, 562, 1969.
29. Kennedy, D. R., Hamilton, T. R., Jensen, J., and Syverton, J. T., Experimental studies of focal infection in monkeys: the systemic effects of hemolytic streptococci in root canals of teeth, *J. Dent. Res.*, 30, 481, 1951.

30. Okada, H., Aono, M., Yoshida, M., Munemoto, K., Nishida, O., and Yokomizo, I., Experimental study of focal infection in rabbits by prolonged sensitization through dental pulp canals, *Arch. Oral Biol.*, 12, 1017, 1967.

31. Allard, U., Nord, C-E., Sjöberg, L., and Strömberg, T., Experimental infections with *Staphyloccus aureus, Streptococcus sanguis, Pseudomonas aerugenosa* and *Bacteroides fragilis* in the jaws of dogs, *Oral Surg. Oral Med. Oral Pathol.*, 48, 454, 1979.

32. Silver, J. G., Martin, L., and McBride, B. C., Recovery and clearance rates of oral microorganisms following experimental bacteraemias in dogs, *Arch. Oral Biol.*, 20, 675, 1975.

33. Bahn, S. L., Goveia, G., Bitterman, P., and Bahn, A. N., Experimental endocarditis induced by dental manipulation and oral streptococci, *Oral Surg. Oral Med. Oral Pathol.*, 45, 549, 1978.

34. Rizzo, A. A., Hampp, E. G., and Mergenhagen, S. E., Spirochetal abscesses in hamster cheek pouch, *Arch. Oral Biol.*, 5, 63, 1961.

35. Hampp, E. G. and Mergenhagen, S. E., Experimental infections with oral spirochetes, *J. Infect. Dis.*, 109, 43, 1961.

36. Swenson, H. M. and Muhler, J. C., Induced fuso-spirochetal infection in dogs, *J. Dent. Res.*, 26, 161, 1947.

37. Rizzo, A. A. and Mergenhagen, S. E., Histopathologic effects of endotoxin injected into rabbit oral mucosa, *Arch. Oral Biol.*, 9, 659, 1964.

38. Taichman, N. S. and Courant, P., The production of haemorrhagic necrosis in the hamster cheek pouch by bacterial endotoxins and catecholamines, *Arch. Oral Biol.*, 10, 541, 1965.

39. Adams, D. and Jones, J. H., Life history of experimentally induced acute oral candidiasis in the rat, *J. Dent. Res.*, 50, 643, 1971.

40. Jones, J. H. and Russell, C., The histology of chronic candidal infection of the rat's tongue, *J. Pathol.*, 113, 97, 1974.

41. Russell, C. and Jones, J. H., The histology of prolonged candidal infection in the rat's tongue, *J. Oral Pathol.*, 4, 330, 1975.

42. Jones, J. H. and Russell, C., Experimental oral candidiasis in weanling rats, *J. Dent. Res.*, 52, 182, 1973.

43. Buditz-Jorgensen, E., Immune response to *C. albicans* in monkeys with experimental candidiasis in the palate, *Scand. J. Dent. Res.*, 81, 360, 1973.

44. Rizzo, A. A. and Ashe, W. K., Experimental herpetic ulcers in rabbit oral mucosa, *Arch. Oral Biol.*, 9, 713, 1964.

45. Henson, D. and Strano, A. J., Mouse cytomegalovirus. Necrosis of infected and morphologically normal submaxillary gland acinar cells during termination of chronic infection, *Am. J. Pathol.*, 68, 183, 1972.

46. Joseph, C. E., Grand, N. G., and Pumper, R. W., An ultrastructural study of the formation of "dark" and "light" cytoplasmic inclusions by murine cytomegalovirus in mouse submandibular gland, *J. Dent. Res.*, 57, 91, 1978.

47. Kreshover, S. J. and Hancock, J. A., Jr., The pathogenesis of abnormal enamel formation in rabbits inoculated with vaccinia, *J. Dent. Res.*, 35, 685, 1956.

48. Kreshover, S. J., Clough, O. W., and Hancock, J. A., Vaccinia infection in pregnant rabbits and its effect on maternal and fetal dental tissues, *J. Am. Dent. Assoc.*, 49, 549, 1954.

49. Kreshover, S. J. and Hancock, J. A., Jr., The effect of lymphocytic choriomeningitis on pregnancy and dental tissues in mice, *J. Dent. Res.*, 35, 467, 1956.

50. Lewin-Epstein, J. and Neder, A., Experimental chronic osteomyelitis in maxillae of rabbits, *J. Dent. Res.*, 43, 456, 1964.

51. Triplett, R. G., Gilmore, J. D., and Armstrong, G. C., An experimental model for osteomyelitis of the mandible, *J. Dent. Res.*, 56, 1520, 1977.

52. Sela, J., Dishon, T., Rosenmann, E., Ulmansky, M., and Boss, J. H., Experimental allergic sialoadenitis. V. Comparison of the response of the parotid gland and synovial membrane to multiple antigenic challenges, *J. Oral Pathol.*, 2, 7, 1973.

53. White, S. C. and Casarett, G. W., Induction of experimental autoallergic sialoadenitis, *J. Immunol.*, 112, 178, 1974.

54. White, S. C., Experimental autoallergic parotitis, *J. Oral Pathol.*, 2, 341, 1973.

55. Boss, J. H., Rosenmann, E., and Sela, J., Experimental allergic sialoadenitis. X. Chronic destructive parotitis induced in immunized rats by daily intraoral challenges with antigens, *J. Oral Pathol.*, 6, 96, 1977.

56. Whaley, K. and MacSween, R. N. M., Experimental induction of immune sialoadenitis in guinea pigs using different adjuvants, *Clin. Exp. Immunol.*, 17, 681, 1974.

57. Whittaker, D. K. and Wilson, T. R., The effect of age and strain differences on the incidence of restraint-induced oral and gastric ulcers in three strains of rats, *J. Dent. Res.*, 51, 619, 1972.

58. Zackin, S. J. and Goldhaber, P., Experimental production of oral ulcerations in the rat, *Oral Surg. Oral Med. Oral Pathol.*, 13, 1267, 1960.

59. Taylor, G. N., Christensen, W. R., Jee, W. S. S., and Rehfield, C. E., Gingival ulceration in beagles induced by Ra²²⁶, Ra²²⁸ and Th²²⁸, *J. Dent. Res.*, 43, 35, 1964.
60. Attström, R. and Egelberg, J., Effect of experimental leukopenia on chronic gingival inflammation in dogs, *J. Periodontal Res.*, 6, 194, 1971.
61. Attström, R., Tynelius-Bratthall, G., and Egelberg, J., Effect of experimental leukopenia on chronic gingivitis in dogs, *J. Periodontal Res.*, 6, 200, 1971.
62. Attström, R. and Schroeder, H. E., Effect of experimental neutropenia on initial gingivitis in dogs, *Scand. J. Dent. Res.*, 87, 7, 1979.
63. Andersen, L., Attstrom, R., and Fejerskov, O., Effect of experimental neutropenia on oral wound healing in guinea pigs, *Scand. J. Dent. Res.*, 86, 237, 1978.
64. Lopez, C. and Miller, A., Investigation of oral changes in experimental polycythaemia, *Arch. Oral Biol.*, 9, 205, 1964.
65. Marefat, P. and Shklar, G., Experimental production of lingual leukoplakia and carcinoma, *Oral Surg. Oral Med. Oral Pathol.*, 44, 578, 1977.
66. Craig, G. E. and Franklin, C. D., The effect of turpentine on hamster cheek pouch mucosa: a model of epithelial hyperplasia and hyperkeratosis, *J. Oral Pathol.*, 6, 268, 1977.
67. Capodanno, J. A. and Hayward, J. R., Effect of keratolytic drugs on oral hyperkeratosis in Syrian hamsters, *J. Dent. Res.*, 45, 951, 1966.
68. Bartlett, P. F., Radden, B. G., and Reade, P. C., The experimental production of odontogenic keratocysts, *J. Oral Pathol.*, 2, 58, 1973.
69. Soskolne, W. A., Bab, J., and Sochat, S., Production of keratinizing cysts within mandibles of rats with autogenous gingival epithelial grafts, *J. Oral Pathol.*, 5, 122, 1976.
70. Riviere, G. R. and Sabet, T. Y., Experimental follicular cysts in mice, *Oral Surg. Oral Med. Oral Pathol.*, 36, 205, 1973.
71. Al-Talabani, N. G. and Smith, C. J., Experimental dentigerous cysts and enamel hypoplasia: their possible significance in explaining the pathogenesis of human dentigerous cysts, *J. Oral Pathol.*, 9, 82, 1980.
72. Burstone, M. S. and Levy, B. M., The production of experimental apical granulomata in the Syrian hamster, *Oral Surg. Oral Med. Oral Pathol.*, 3, 807, 1950.
73. Summers, L. and Papadimitriou, J., The nature of epithelial proliferation in apical granulomas, *J. Oral Pathol.*, 4, 324, 1975.
74. Bhaskar, S. N., Bolden, T. E., and Weinmann, J. P., Pathogenesis of mucoceles, *J. Dent. Res.*, 35, 863, 1956.
75. Chaudry, A. P., Reynolds, D. H., LaChapelle, C. F., and Vickers, R. A., A clinical and experimental study of mucocele (retention cysts), *J. Dent. Res.*, 39, 1253, 1960.
76. Harrison, J. D. and Garrett, J. K., Experimental salivary mucoceles in cats. A histochemical study, *J. Oral Pathol.*, 4, 297, 1975.
77. Bhaskar, S. N., Bolden, T. E., and Weinmann, J. P., Experimental obstructive adenitis in the mouse, *J. Dent. Res.*, 35, 852, 1956.
78. Leake, D. L., Haydon, G. B., and Laub, D., The microscopy of parotid atrophy after ligation of Stensen's duct, *J. Oral Pathol.*, 3, 167, 1974.
79. Standish, S. M. and Shafer, W. G., Serial histologic effects of rats submaxillary and sublingual salivary gland duct and blood vessel ligation, *J. Dent. Res.*, 36, 866, 1957.
80. Bhaskar, S. N., Lilly, G. E., and Bhussry, B., Regeneration of the salivary glands in the rabbit, *J. Dent. Res.*, 45, 37, 1966.
81. Castelli, W. A., Nasjleti, C. E., and Diaz-Perez, R., Interruption of the arterial alveolar flow and its effects on the mandibular collateral circulation and dental tissues, *J. Dent. Res.*, 54, 708, 1975.
82. Dreizen, S., Stone, R. E., and Stahl, S. S., The effect of experimentally induced generalized medial arteriosclerosis on the oral vessels of the rat, *Arch. Oral Biol.*, 8, 187, 1963.
83. Bhussry, B. R. and Rao, S., Effect of sodium diphenylhydantoin on oral mucosa of rats, *Proc. Soc. Exp. Biol. Med.*, 113, 595, 1963.
84. Staple, P. H., Reed, M. J., and Mashimo, P. A., Diphenylhydantoin gingival hyperplasia in *Macaca arctoides*: a new human model, *J. Periodontol.*, 48, 325, 1977.
85. Owen, L. N., The effects of administering tetracyclines to young dogs with particular reference to localization of the drugs in the teeth, *Arch. Oral Biol.*, 8, 715, 1963.
86. Monto, R. W., Fine, G. F., and Rizek, R. A., Exfoliative cells of the oral mucous membranes in patients receiving chemotherapy for malignant disease, *J. Oral Surg. Anesth. Hosp. Dent. Serv.*, 21, 95, 1963.
87. Ghee, H. T., Han, S. S., and Avery, J. K., A study of salivary glands of rats injected with Actino-mycin D, *Am. J. Anat.*, 116, 631, 1965.
88. Adkins, K. F., Ultrastructural changes induced by Actinomycin D in acinar cells in submandibular glands of rats, *Arch. Oral Biol.*, 19, 859, 1974.

89. Stene, T., Vincristine's effect on dentinogenesis in rat incisor, *Scand. J. Dent. Res.*, 87, 39, 1979.

90. Yen, P. K. and Shaw, J. H., Effects of repeated ingestion of large quantities of acetylsalicylic acid on membranous bone growth and dentin apposition in young monkeys, *J. Dent. Res.*, 56, 1265, 1977.

91. Butcher, E. O. and Taylor, A. C., The effects of denervation and ischemia upon the teeth of the monkey, *J. Dent. Res.*, 30, 265, 1951.

92. Guth, L., The effects of glossopharyngeal nerve transection on the circumvallate papilla of the rat, *Anat. Rec.*, 128, 715, 1957.

93. Boyd, T. G., Castelli, W. A., and Huelke, D. F., Removal of the temporalis muscle from its origin: effects on the size and shape of the coronoid process, *J. Dent. Res.*, 46, 997, 1967.

94. Chaudry, A. P., Schwartz, D., Schwartz, S., and Schmutz, J. A., Jr., Some aspects of experimental induction of cleft palates, *J. Oral Ther. Pharmacol.*, 4, 98, 1967.

95. Blaustein, F. M., Feller, R., and Rosenzweig, S., Effect of ACTH and adrenal hormones on cleft palate frequency in CD-1 mice, *J. Dent. Res.*, 50, 609, 1971.

96. Metah, D., Reznik, G., and Schlegel, D., Experimentally induced cleft palates and prophylactic studies in Swiss mice, *J. Oral Pathol.*, 5, 209, 1976.

97. Medak, H. and Burnett, G. W., The effect of X-ray irradiation on the oral tissues of the Macacus Rhesus monkey, *Oral Surg. Oral Med. Oral Pathol.*, 7, 778, 1954.

98. Burstone, M. S., Histological changes in the oral tissues in response to continuous irradiation, *J. Exp. Zool.*, 115, 341, 1950.

99. Liu, H. M., Meyer, J., and Waterhouse, J. P., An ultrastructural study on the effects of X-irradiation on the oral epithelium of the rat, *J. Oral Pathol.*, 5, 194, 1976.

100. Sholley, M. M., Sodicoff, M., and Pratt, N. E., Early radiation injury in the rat parotid gland, *Lab. Invest.*, 31, 340, 1974.

101. Shafer, W. G., The effect of single and fractionated doses of selective applied X-ray irradiation on the histologic structure of the major salivary glands of the rat, *J. Dent. Res.*, 32, 796, 1953.

102. English, J. A., Wheatcroft, M. G., Lyon, H. W., and Miller, C., Long-term observations of radiation changes in salivary glands and the general effects of 1000 r to 1750 r of X-ray radiation locally administered to the heads of dogs, *Oral Surg. Oral Med. Oral Pathol.*, 8, 87, 1955.

103. Mayo, J., Carranza, F. A., Jr., Epper, C. E., and Cabrini, R. L., The effect of total-body irradiation on the oral tissues of the Syrian hamster, *Oral Surg. Oral Med. Oral Pathol.*, 15, 739, 1962.

104. Rohrer, M. D., Kim, Y., and Fayos, J. V., The effect of cobalt[60] irradiation on monkey mandibles, *Oral Surg. Oral Med. Oral Pathol.*, 48, 424, 1979.

105. Meyer, I., Shklar, G., and Turner, J., A comparison of the effects of 200 KV radiation and cobalt-60 radiation on the jaws and dental structure of the white rat, *Oral Surg. Oral Med. Oral Pathol.*, 15, 1098, 1962.

106. Furstman, L. L., Effect of X-irradiation on the mandibular condyle, *J. Dent. Res.*, 49, 419, 1970.

107. Burstone, M. S., The effect of X-ray irradiation on the development of the mandibular joint of the mouse, *J. Dent. Res.*, 29, 358, 1950.

108. Brunst, V. V., Sheremetieva-Brunst, E. A., and Figge, F. H. J., The effect of local X-ray irradiation upon the teeth and surrounding tissues in young axolotls *(Siredon mexicanum)*, *J. Dent. Res.*, 31, 609, 1952.

109. Burstone, M. S., The effect of X-ray irradiation on the teeth and supporting structures of the mouse, *J. Dent. Res.*, 29, 220, 1950.

110. Medak, H., Weinreb, W., Sicher, H., Weinmann, J. P., and Schour, I., The effect of single doses of irradiation upon the tissues of the upper rat incisor, *J. Dent. Res.*, 31, 559, 1952.

111. Medak, H., Schour, I., and Klauber, W. A., Jr., The effect of single doses of irradiation upon the eruption of the upper rat incisor, *J. Dent. Res.*, 29, 839, 1950.

112. English, J. A., Schlack, C. A., and Ellinger, F., Oral manifestations of ionizing radiation. II. Effect of 200 KV X-ray on rat incisor teeth when administered locally to the head in the 1500 r dose range, *J. Dent. Res.*, 33, 377, 1954.

113. Medak, H., Ortel, J. S., and Burnett, G. W., The effect of X-ray irradiation on the incisors of the Syrian hamster, *Oral Surg. Oral Med. Oral Pathol.*, 7, 1011, 1954.

114. Kaplan, W. V. and Bruce, K. W., Effects of irradiation on the forming dentition of ten-day-old Syrian hamsters, *Oral Surg. Oral Med. Oral Pathol.*, 6, 1348, 1953.

115. English, J. A. and Tullis, J. L., Oral manifestations of ionizing radiation. I. Oral lesions and effect on developing teeth of swine exposed to 2000 KV total body X-ray irradiation, *J. Dent. Res.*, 30, 33, 1951.

116. Cowgiel, J. W., Eruption of irradiation-produced rootless teeth in monkeys, *J. Dent. Res.*, 40, 538, 1961.

117. Pratt, N. E., Sodicoff, M., Liss, J., Davis, M., and Sinesi, M., Radioprotection of the rat parotid gland by WR-2721: morphology at 60 days postirradiation, *Int. J. Rad. Oncol. Biol. Phys.*, 6, 431, 1980.

118. DeCoursey, E., Atomic bomb radiation and the oral regions, *Oral Surg. Oral Med. Oral Pathol.,* 5, 179, 1952.
119. Cowgiel, J. M., Experimental radio-osteonecrosis of the jaws, *J. Dent. Res.,* 39, 176, 1960.
120. Bond, W. R., Jr., Matthews, J. L., and Finney, J. W., The influence of regional oxygenation on osteoradionecrosis, *Oral Surg. Oral Med. Oral Pathol.,* 23, 99, 1967.
121. Shteyer, A. and Howell, R. M., Tissue reactions to chronic irritation of the tongue in germfree and conventional rats, *J. Oral Surg.,* 28, 109, 1970.
122. Bhaskar, S. N. and Lilly, G. E., Traumatic granuloma of the tongue (human and experimental), *Oral Surg. Oral Med. Oral Pathol.,* 18, 206, 1964.
123. Chierici, G., Harvold, E. P., and Dawson, W. J., Primate experiments in facial asymmetry, *J. Dent. Res.,* 49, 847, 1970.
124. Taylor, R., Shklar, G., and Roeber, F., The effects of laser radiation on teeth, dental pulp and oral mucosa of experimental animals, *Oral Surg. Oral Med. Oral Pathol.,* 19, 786, 1965.
125. Parker, W. S., Frisbe, H. E., and Grant, T. S., The experimental production of dental ankylosis, *Angle Orthodont.,* 34, 103, 1964.
126. Butcher, E. O. and Klingsberg, J., Age, gonadectomy, and wound healing in the palatal mucosa of the rat, *Oral Surg. Oral Med. Oral Pathol.,* 16, 484, 1963.
127. Irving, J. T., Experimental enamel hypoplasia in rats, *Br. J. Pathol.,* 31, 458, 1950.
128. Kreshover, S. J., Pathogenesis of enamel hypoplasia: an experimental study, *J. Dent. Res.,* 23, 231, 1944.
129. Farbman, A. I., Differentiation of oral epithelium and taste bud development, *J. Dent. Res.,* 50, 1422, 1971.
130. Trefz, B., Histochemical investigation of the modal specificity of taste, *J. Dent. Res.,* 51, 1203, 1972.
131. Miller, N. C., Bishop, J. G., and Dorman, H. L., Activation of medullary neurons by noxious stimuli applied to the teeth, *J. Dent. Res.,* 51, 577, 1972.
132. Spouge, J. D. and Cutler, B. S., Hypersensitivity reactions in mucous membranes. II. An investigation into induced hypersensitivity lesions in the mucous membranes of laboratory animals, *Oral Surg. Oral Med. Oral Pathol.,* 16, 539, 1963.
133. Cooke, W., Jr., Experimental production of micrognathia and glossoptosis associated with cleft palate (Pierre Robin syndrome), *Plast. Reconstr. Surg.,* 38, 395, 1966.
134. Vickers, R. A. and von der Muhl, C. H., An investigation concerning inducibility of lymphoepithelial cysts in hamsters by autologous epithelial transplantation, *J. Dent. Res.,* 45, 1029, 1966.
135. Hals, E. and Nielsen, K., Dental changes in experimental bile duct obstruction in rodents, *Arch. Oral Biol.,* 14, 151, 1969.
136. Aleves, J. B. and Machado, C. R. S., Histological and histoquantitative study of the rat submandibular gland in Chagas disease, *Arch. Oral Biol.,* 25, 437, 1980.

EXPERIMENTAL ORAL CARCINOGENESIS

INTRODUCTION

How common is oral cancer? Whom does this disease affect? What are its consequences? How important — in terms of morbidity and mortality, in terms of human suffering — is oral cancer? Why should we spend time, money, and human energy to study the etiology, pathogenesis, prevention, and treatment of oral cancer? The data necessary to answer such questions are generated primarily through epidemiological studies.

Epidemiology of Oral Cancer

Oral cancer accounts for about 3 to 5% of all cancers in most of the western countries, a sharp contrast with the 40% frequency rate of oral cancer in some parts of India. As Pindborg[1] pointed out, comparisons between various population groups are difficult and frequently misleading. The 40% figure for the relative frequency of oral cancer in some areas of India was derived from data obtained in cancer hospitals and through cancer departments of teaching and research institutions. In some countries, frequency rates are determined through evaluation of data obtained from cancer registries and from biopsy reports in oral pathology departments of various schools and treatment centers. A cancer hospital with a well-recognized team of oral cancer therapists might well present figures which have no relationship to the actual frequency rate in the general population. For example, oral cancer accounted for some 20% of the admissions to the Sloan Kettering Memorial Hospital in New York in 1950.[2] One cannot, however, infer from those figures that oral cancer accounts for 20% of all malignant disease in the New York City area. What we can infer is that there was a team of specialists interested and expert in the treatment and rehabilitation of the oral cancer patient.

The incidence of oral cancer should be established by using national or regional cancer registries and/or by careful investigation of particular well-defined populations during an established time period. The International Agency for Research on Cancer, in collaboration with the International Union Against Cancer and the International Association of Cancer Registries, has published incidence figures for cancer on a global basis since 1966. Reliable statistics are that in all countries, men have a higher rate of oral cancer than women, but there are marked variations in the male to female ratio in the world, the lowest being found in the Scandinavian countries.

Striking differences occur in the incidence of oral cancer on the European continent. Oral cancer is twice as frequent in rural as in city areas. Malta has almost seven times more oral cancer than the south metropolitan region of the United Kingdom. The explanation for the high incidence in Malta and rural areas of Europe is the greater facial exposure to sunlight in those areas, which probably accounts for the high number of lip cancers. There is also a high incidence of oral cancer in Hungary and in Spain as compared to the occurrence in other European countries, but this difference has not been explained.

In the Americas the differences between and within countries are more pronounced than in Europe. The highest incidence of oral cancer in the western part of the world was found in Newfoundland, where there are some 30 cases per 100,000 population annually. The high rate may be correlated with the fact that fishermen in Newfoundland have a probability of developing lip cancer approximately 4.5 times as great as that of other men of comparable age in Newfoundland.[3] The high rate of lip cancer found in other provinces of Canada cannot be explained.

In the U.S. the incidence rate of oral cancer varies from state to state and even within the same state. Whites are more often affected than blacks, probably because the incidence figures include lip cancer, which is quite low in the black population.

While it has been known for many years that some countries in Southeast Asia have an extremely high frequency of oral cancer, only in recent years have careful epidemiological studies provided hard data to support that finding. In 1966 Hirayama[4] visited six countries in Southeast Asia and produced evidence for the association of oral cancer with tobacco chewing. He also found an association, although less marked, between smoking, drinking alcohol, and eating a vegetarian diet with oral cancer.

Since oral cancer is frequently fatal, the number of deaths from oral cancer is an indication of the occurrence of the disease. World health statistics[5] provide annual death rates from malignant neoplasms of the buccal cavity and pharynx. The highest mortality rate was 11.8/100,000, found in Hong Kong. It probably reflects the large number of nasopharyngeal cancers in the Chinese population there. Outside of Asia, the highest mortality rate for oral cancer was 9.5 deaths per 100,000, reported from France.

In the last few years there has been a tendency for oral cancer mortality rates to decrease. This finding is in contrast to the increased mortality for cancer of some other sites. Russells[6] reported that the mortality rate for oral cancer in men decreased by 40% in the 5-year period from 1945 to 1949 as compared with the period from 1935 to 1939, but that the mortality rate for oral cancer in women increased by 20% for the same time periods. Later reports from England and from Canada confirmed the decreasing male mortality rate, whereas the mortality rate for women appears to have remained unchanged.[7,8] A significant decline in the incidence of oral cancer has occurred in Australia, Germany, Czechoslovakia, and the U.S., especially Texas. In other countries, such as Denmark and Finland, there has been no change in oral cancer morbidity.

Oral cancer is a disease which characteristically occurs in men of advanced age. For example, among Caucasians, nine of ten lip cancers occur in men at least 45 years old, and half occur in men over 64 years old.[9] The prognosis of oral cancer (apart from the lip) is rather poor, probably because there is often delayed diagnosis and treatment. The relative ease of detection of oral cancer should result in early diagnosis and high cure rates. Most oral cancers, however, are not recognized at a time when satisfactory treatment can be accomplished. The 5-year survival rate of patients with localized lesions is 70%, whereas the 5-year survival rate of patients with lesions which have spread beyond the local origin is only 24%.[10] The incidence of cervical metastases increases markedly as the size of the tumor increases. Thus, there is ample evidence of the value of early treatment of intraoral cancer, which should provide both incentive and challenge to dentists to seek out cancer of the mouth in its early stages. The reasons for delay in diagnosis have been analyzed by Cooke.[11] He points out that early oral cancer is usually painless and does not interfere with the functions of the mouth. The elderly and frail do not want to be troubled by visits to doctors or to hospitals. He also indicated that delay occurred when malignant disease was not suspected and the lesion was treated as a fungus or other infection with antifungus drugs and antibiotics, steroids, or mouthwashes. He indicated that the general medical practitioner was more likely to delay biopsy and thus diagnosis and treatment than the general dental practitioner. It seems, therefore, that the discovery of malignant lesions is the responsibility of dentists who are consulted at regular intervals for dental examinations.

Although epidemiology is not an "experimental science", epidemiological studies of oral cancer may provide data which can lead to an understanding of its etiology. Extensive epidemiological studies, especially in Southeast Asia, have convincingly demonstrated the correlation between tobacco usage and oral carcinomas. The need for continuing epidemiological studies, with special attention to the possible etiology of malignant and premalignant oral lesions, is obviously a fertile field for continued research. An innovative and imaginative approach to such studies is the challenge.

EXPERIMENTAL CARCINOGENESIS

The fundamental aspects of experimental carcinogenesis must be elaborated if we are to attempt to understand the problems of oral cancer. Cancer research per se is less than 100 years old, but experimental studies during the past 80 years have resulted in a literature so vast that a comprehensive review is almost impossible. Cowdry[12] pointed out that the history of cancer research cannot be taken up in an orderly fashion because many specialists in the field of the biological and physical sciences have attacked various aspects of the cancer problem from within their own special sphere of interest. Thus, although great progress has been made by geneticists, pathologists, chemists, virologists, endocrinologists, physicists, behavioral scientists, and others, each adding to a better understanding of the problems of growth in general and of malignant growth in particular, a sequential review of the literature is impossible. It is, however, important to present some background material in the general area of carcinogenesis so that the field of experimental oral carcinogenesis can be viewed with some perspective.

At the end of the last century, spontaneous tumors in animals were dismissed as unrelated to malignant disease in man. It soon became obvious, however, that the study of spontaneous tumors in animals could substantially add to our knowledge of tumor pathogenesis; so the search for an appropriate animal model began. Epidemiologic studies with animals revealed that 10% of dogs, 0.5% of cats, and 2% of horses developed spontaneous tumors. The incidence of spontaneous tumors in fowl was much higher. Nevertheless, researchers decided, for very practical reasons, that mice would be the most appropriate animal in which to study spontaneous tumors. Large groups of mice could be maintained in relatively small quarters.

Investigators soon discovered that groups of mice housed in small cages seem to develop more tumors in some cages than in others. It was first thought that spontaneous tumors in mice were an infectious disease. Many bacteriological studies of mouse tumors and of mouse-bearing tumors were reported. It was soon learned that the disease was not infectious but was, in fact, hereditary and that the reason why animals housed together in certain cages had more tumors than others was that they had a common genetic pool. Studies of the hereditary aspects of malignant disease in mice were hampered by the fact that mouse stocks were quite heterogeneous. Such geneticists as Little, Lynch, Bagg, Strong, Slye, and others responded to the need for a homozygous strain of mouse. They proposed that 20 inbred generations of mice be called a "pure strain", and many laboratories proceeded to produce such strains. Today there are over 80 strains of mice, and inbred mice are considered as necessary to biological work as pure chemicals are to chemical investigation.

Much important information came from the development of inbred strains of mice. The use of inbred mice permitted the observation of large numbers of genetically identical animals which developed almost every type of neoplasm encountered in man. The inbreeding apparently did not influence the development of neoplasms other than by concentrating a particular type of tumor within a strain by segregation. For example, analysis of the genetic factors involved in the susceptibility to mammary tumors was made by means of rather ingenious hybridization experiments with homozygous mice. It was shown early that hybrid females resulting from the mating of high tumor strain females to low tumor strain males developed mammary tumors in approximately the same incidence as the high tumor strain. On the other hand, when the reciprocal cross was made, that is, when the low tumor strain females were mated with high tumor strain males, the tumor incidence in the hybrid offspring was about the same as for the low tumor strain. It was obvious that an extrachromosomal factor was involved. The factor could be transmitted through either the cytoplasm of the egg, the placenta, or the milk. Bittner[13] undertook the investigation of the third possibility and, by a remarkably simple procedure, opened a new chapter on the problem of mammary tumors, especially in mice. He foster-nursed low tumor strain newborn animals to high tumor strain mothers and high tumor strain newborn animals to low tumor strain mothers. The animals took on the incidence of mammary tumors of the foster mother. He thus found that a factor in the milk (since shown to be a virus) was responsible for the transmission of the high tumor strain mammary cancer characteristic. While this work was going on, a considerable amount of research on the endocrine interrelationships of mammary cancer was done. Virgin females of the high tumor strain mouse, for example, were shown to develop fewer mammary cancers than did breeding females of the same strain. Ovariectomized mice developed fewer mammary cancers than did virgin females. Males, which normally develop few mammary cancers, developed some mammary neoplasms following castration, but developed a considerable number following castration and estrogen injection.

Chemical Carcinogenesis

Shimkin[14] designated those agents capable of inducing neoplasms in animals as "carcinogenic agents". These agents might be either chemical, physical, or living substances or combinations of any of these three. As we shall point out later, epidemiological studies, as well as experimental animal studies, provided evidence that a number of etiologic factors are implicated in various oral cancers. A direct causative relationship between the agent or agents employed and the neoplasm produced is not implied. All that can be said is that in animals that are injected with certain agents or exposed to certain procedures, neoplasms arise in significantly higher incidence than in untreated animals. The list of chemicals in the environment capable of producing malignant disease seems interminable.

Polycyclic Hydrocarbons

Since the time of Sir Percival Pott,[15] who showed that carcinoma of the scrotum in chimney sweeps was related to the embedding of soot in the scrotal folds, a search for carcinogenic hydrocarbons has been made. Almost 100 years later, von Volkman[16] implicated coal tar as a cause of industrial skin cancer, and Bell[16] discovered "paraffin cancer" due to the contact with shale oil.

In 1887 "mule spinner's cancer" among the cotton spinners of Lancashire was shown to be associated with exposure to mineral oil. It was not until 1915, however, that Yamagiwa and Ichikawa[17] induced skin tumors in the ears of rabbits by the continuous application of coal tar, and Tsutsui[16] shortly thereafter produced skin cancers in mice by painting them with coal tar. Their findings were reported some 140 years after the discovery by Sir Percival Pott that the soot in chimneys was responsible for skin tumors and some 40 years after von Volkman implicated coal tar as an etiologic agent in skin cancer. Moreover, it was not until the early 1930s that Kennaway and Heiger[18] isolated the polycyclic hydrocarbons from coal tar and showed them to be carcinogens. In 1930, 1,2,5,6-dibenzathracene was isolated, and later 3,4-benzpyrene was described.

A special interest in the carcinogenic hydrocarbons was aroused when one of the most active polycyclic hydrocarbons, 20-methylcholanthrene, was synthesized from bile acids. Because the structural resemblance among the carcinogenic hydrocarbons to cholesterol, bile acids, and steroid hormones was so obvious, hopes were high that a common molecular structure, one that was elaborated by the body, would clarify the cancer problem. Thus, the polycyclic hydrocarbons have been intensively studied for their carcinogenic activity by many workers. They are known as "universal carcinogens" because they induce malignant disease after topical application, after injection s.c., after injection i.m or i.v., or after feeding. Because most of the work in experimental carcinogenesis has been done on skin, most is known about skin cancer.

EXPERIMENTAL ORAL CARCINOGENESIS

Tobacco

As noted above in the discussion of the epidemiology of oral cancer, reference is constantly made to the importance of tobacco as an etiologic factor in the development of oral malignancies. Most of the hard data implicates snuff-dipping, betel nut chewing with lime and tobacco, and reverse smoking. Cigar, cigarette, and pipe smoking have also been associated with oral cancer.

Most of the studies dealing with the relationship of tobacco smoking and oral cancer were conducted by Wynder and co-workers.[19,20] Wynder was able to demonstrate a significant dose-response relationship for both cigar and pipe smoking with oral cancer development. Graham et al.[21] demonstrated that patients who smoked heavily, that is, more than one package of cigarettes per day or more than five cigars or pipe bowls per day, had a risk of developing oral cancer six times that of individuals who did not smoke. On the other hand, Jafarey and Zaidi[22] pointed out that in Pakistan, combining smoking with the chewing of betel nut and tobacco increased the risk of developing oral cancer by 23 times in men and 35 times in women. A very careful study of tobacco-induced epithelial proliferation in the palate of human subjects was made by Chapman and Redish.[23] In the course of a dental survey of elderly patients in their hospital, they found 15 with varying degrees of leukoplakia of the hard palate. These individuals were questioned in depth, and quantitative and qualitative data on smoking habits were obtained. Since the researchers were unable to find pertinent information as to the temperature of tobacco smoke issuing from the bit-end of a pipe when smoked in the usual manner, they developed their own technique for measuring the smoke as it reached the palate. They ingeniously affixed a thermometer to the pipe so that it just touched the palatal mucosa when the pipe was smoked in the usual manner. For most patients the temperature ranged between 98.4° and 99.8°F, although in one patient the temperature reached 100.4°F and in two others 102°F and 102.4°F. Lesions were graded both clinically and histologically. Many of the individuals in the study smoked for more than 50 years. One smoked for only 3 years; several, for 20 years. Except for the one patient who smoked only 3 years, none of the patients smoked for less than 20 years. The study demonstrated that the severity of the epithelial proliferation was related to the length of time the patient smoked and that the heat of the pipe tobacco smoke and the total accumulated amount of tobacco smoked were not factors in the degree of leukoplakia. The study implies that there is a cumulative time effect of tobacco smoke on epithelial proliferation and that the palatal mucosa might be an excellent site to study the early cytological effects of tobacco smoke on human mucosal epithelium.

Many authors have pointed out that "betel-chewing" is used as a generic term to describe the chewing of various concoctions. The makeup of the betel quid is largely determined by local custom. The quid usually contains, in addition to betel nut, tobacco and/or lime, and, occasionally, various spices. Hirayama[24] pointed out that oral cancer is infrequent in Afghanistan and Nigeria, where tobacco is chewed without lime, but is common in New Guinea, where betel nut and lime are chewed without tobacco. The observation seems to incriminate lime rather than tobacco as a carcinogenic substance in the betel quid. The evidence from epidemiological studies produced many attempts to investigate the carcinogenic potential of the various ingredients in the betel quid and/or snuff through tests on experimental animals.

Since the work in 1954 of Salley,[25] who demonstrated that malignant tumors could be produced in the cheek pouch of the Syrian hamster by regular treatment with a known chemical carcinogen, the hamster cheek pouch mucosa has been used as a test organ for the exploration of possible carcinogenic agents. The hamster cheek pouch is indeed a fascinating tissue in which to explore for possible carcinogenic agents and does provide an excellent model for that purpose. There is, however, considerable doubt as to whether the hamster cheek pouch should be considered "oral mucosa". It seems to be a very special organ and will therefore be discussed in considerable detail later in this chapter. Suffice to say here that when the hamster cheek pouch was treated with slaked lime as calcium hydroxide,[26] epithelial atypia developed. Snuff, chewing tobacco, and cigarette smoke tar did not produce neoplasms in hamster cheek pouches.[27]

A review of attempts to produce experimental tumors in monkeys[28] reveals that monkeys appear to exhibit great resistance to carcinogens in all anatomical sites. While this resistance may help to explain the apparent low incidence of spontaneous malignant disease in monkeys, the fact that relatively few monkeys in captivity live to an advanced age should also be considered. The few recorded cases of oral cancer in monkeys occurred in very old animals.

In an effort to study the histopathogenesis of malignant disease in the oral tissues of monkeys and to provide diagnostic criteria for the histological assessment of precancerous oral lesions, Cohen and Smith[29] made an attempt to produce epithelial changes in the mucosa of the monkey cheek pouch, which they isolated surgically from continuity with the oral cavity. Application of a potent chemical carcinogen to the oral mucosa of monkeys at biweekly intervals over long periods of time failed to produce a neoplastic response. The determination of whether the oral mucosa possessed resistance to the action of a carcinogen or whether the negative results were attributable to either dilution or dispersion by saliva was important. It was also important to investigate the effects on the oral mucosa of various components in chewing tobacco. A sealed system whereby the substances to be tested would be kept in contact with the mucosa for long periods of time was devised in order to provide the necessary environment for testing potential carcinogenic agents without disturbance. Adult *Macaca irus* were the monkeys chosen for the experiment. Two substances were tested: (1) 7,12-dimethylbenzanthracene (DMBA) as a 0.5% solution in acetone and in liquid paraffin, and (2) the tobacco constituent of the betel quid commonly chewed in the Uttar Pradesh region of India, referred to as Mainpuri tobacco. (The predominant component of the quid is probably tobacco grown in the Mainpuri area. There is also an odor suggestive of oil of cloves or a similar spice associated with the quid. The quid contains an undetermined quantity of lime, the pH of a paste compounded from the quid being approximately 9.0).

Two monkeys, age 3½ and 5 years at the start of the experiment, were anesthetized. Their oral mucosa was examined, and sites were selected for the carcinogen treatment. One monkey received DMBA applications for 15 weeks and the other for 3 years and 10 months at the time of the report. Different sites within the oral cavity were selected for the painting, which included the cheek pouch mucosa, buccal alveolar mucosa, and labial mucosa. A control monkey was painted with acetone for a period of 5 weeks.

In order to be sure that the tobacco and the DMBA would remain in contact with the oral mucosa for an extended period of time, a method was devised by which a portion of the cheek pouch was freed from continuity with the oral mucosa and closed off. Since it was difficult to locate the pouch from an extraoral approach without a reasonably large incision, a marker was enclosed in the pouch in the form of a ball of medical grade fine sponge silicone rubber (Silastic®). When healing was completed, the soft ball of Silastic® demarcated the pouch so unmistakably that the removal of pouch tissue for biopsy examination, as well as the insertion of tobacco paste or other materials to be investigated, could be effected expeditiously with only minor surgical procedures. An electrical tattooing needle was used to augment the application of the carcinogen during the first 15 weeks of the experiment for the purpose of facilitating the penetration of the polycyclic hydrocarbon.

Throughout the course of the experiment utilizing the known carcinogen, no epithelial atypia or recognized premalignant changes were found. Biopsies had been taken early in the experiment and 3½ years after the beginning of the experiment. Histologically, the cheek pouch of DMBA-treated animals exhibited epithelial hyperplasia with a thick, loose parakeratotic surface. Occasional cells exhibited "ballooning" of the cytoplasm.

A group of animals had pellets of tobacco paste inserted through an amalgam carrier in order to avoid contaminating the incision surface. Biopsies were taken from pouches containing this tobacco paste from the seventh postoperative day onwards. The predominant change seen in the early stages of the experiment was a true epithelial hyperplasia with an apparent increase in the cell population of the basal layer. Large areas were present in which the cells appeared to be enormously swollen, although there was no increase in the size of their nuclei, and there was a marked decrease in the affinity of the cytoplasm for the stain. These "balloon-like cells" were instantly recognizable in every biopsy that was taken from the 10th postoperative day onwards. They were frequently found in areas immediately below the epithelial surface as well as in the spinous layer. In animals in which the pouch remained sealed for 30 days after the insertion of the tobacco paste, little more than the basal layer of the epithelium remained free of the strange transformation to balloonlike cells.

The one distinctive feature of the reaction of epithelial cells to the Mainpuri tobacco paste was the ballooning of cells, the microscopic appearance of which was indistinguishable from those described in human oral epithelium exposed over long periods of time to the action of snuff[30] and chewing tobacco.[31] Neither the exact nature of this abnormality nor its prognostic significance is understood. It should be pointed out, however, that the intensity of the effect exerted by the tobacco paste is much greater than that exerted by the powerful carcinogenic agent DMBA.

A recurring problem in the literature on betel chewing arises from the heterogeneous nature of the ingredients included under this generic term. One does not know whether the effects observed are due to the betel leaf, the tobacco, the lime, and/or to the other spices added to the concoction. Since the standard betel quid known in Indian as "pan" is composed of betel leaf, areca palm nut, lime, and catechu — to which tobacco is frequently added — there are five ingredients which may play some role in cancer induction either directly as a carcinogenic agent or indirectly as a cocarcinogen. Hamner, working at the Southwest Foundation for Research and Education in San Antonio, Tex., undertook a long-range study to test the carcinogenic capacity of betal quid ingredients by chronic exposure of the baboon buccal mucosa to those substances.[32-35]

Twelve adult baboons (*Papio cynocephalus*) were conditioned to a deficient dietary protein level of 7.2% in an effort to emulate Indian dietary standards. Although the baboons' weight dropped slightly in the initial phase of the conditioning program, their weight was maintained thereafter on an even scale. The average serum protein level was lowered from a normal of 7.4 to 6.4g% and was maintained at the reduced level over a 4-year period. In order to closely duplicate the "pan" habit, fresh betel leaves were routinely flown by airmail special delivery from Bombay, India, to San Antonio, and kept under moist refrigeration. A betel quid was prepared as a ground mixture of these fresh betel leaves, areca palm nut, and USP calcium hydroxide. Five of the baboons had approximately 3½ g of the basic betal quid placed against the right buccal mucosa three times a week via a plastic open-end syringe. The mucosa had been modified to create a retention pouch which could be sutured and closed. Seven of the animals received the basic betel quid plus Maharashtran sun-dried tobacco. The twelve animals were maintained for 4 years, after which half were placed on a normal diet and the betel quid treatment ceased. Biopsies were taken after 1, 6, 9, 12, 16, 23, 29, 34, 42, 48, and 50 months following the insertion of the quid.

Biopsies that were taken during the first 6 months of either treatment revealed ulceration, inflammation, and healing in the later months. By the 9th month following the insertion of the basic betel quid, there was epithelial hyperplasia with areas of squamous atrophy where scarring had ocurred. Increased parakeratosis was common. After 12 months of treatment, the histologic appearance was similar in all of the five baboons which had received only the basic betel quid. The basal cell nuclei showed more spindling and were hyperchromatic. More inflammatory cells were present than in the earlier biopsies. By the end of 30 months, all of the baboons in the basic betel quid group demonstrated ulcerative hemorrhagic lesions in the buccal pouch, but none exhibited malignant change.

The pouch specimens from baboons which received tobacco mixed with the basic betel quid also demonstrated epithelial hyperplasia, areas of flat, healed-over epithelium, excessive parakeratosis, and chronic inflammation after 9 months of treatment. However, in this group of seven baboons, one exhibited pseudoepitheliomatous hyperplasia and nuclear hyperchromatism. By 12 months, biopsy specimens demonstrated a greater degree of epithelial hyperplasia and increased parakeratosis, intracellular edema, and vacuolation plus a very active "restless" basal cell layer. One of the seven baboons in the combination group presented with obvious epithelial atypia in which the basal and lower spinous cells contained marked hyperchromatic bizarre-shaped nuclei. By 30 months, all of the animals in the combination group had chronic ulcerative hemorrhagic lesions in the treated buccal pouch. The 34-month histologic specimens of the combination group revealed marked epithelial atypia in all pouch specimens. One of the seven baboons exhibited carcinoma *in situ* with loss of the basement membrane and initial downward malignant spread.

The results of these experiments indicated that the local effects of calcium hydroxide served to prepare or condition the buccal mucosa for the enhancement of carcinoma evolvement, using tobacco as the weak carcinogen. The severe caustic action of calcium hydroxide affected both the epithelium and the underlying connective tissue. The covering epithelium was initially lost by ulceration, and the subjacent connective tissue became markedly inflamed. Fibroblastic proliferation, foreign body giant cell formation, and capillary budding occurred as granulation healing transpired in the damaged subepithelial area. The tobacco exerted a definite carcinogenic effect on the epithelial tissues, already damaged and rendered hyperplastic or atrophic by calcium hydroxide. The hypothesis that these two substances do possess the capacity to act together in a cocarcinogenic fashion was substantiated by the findings. Hamner pointed out that the experimental evidence supported the concept of a complex etiology for cancer rather than multiple etiologies. He claimed that such agents as Indian tobacco, along with the proper accessory factors such as calcium hydroxide or protein deficiency, acted in an appropriate combination or sequence to lead to malignant transformation.

In a later study, in which baboons were maintained for longer periods of time, the histologic appearance of the betel quid plus tobacco buccal mucosa showed, in addition to the epithelial atypia, "ballooning" of the cells and cellular proliferation. The ballooning of cells described in these animals was similar to that described by Cohen and his co-workers[29] and closely resembed the changes noted in human leukoplakia and carcinoma *in situ* induced by snuff or tobacco chewing.[1] Biopsies of animals which had betel quid plus tobacco treatment for 42 to 48 months revealed characteristic carcinoma *in situ*. The suggestion was made that by continuing a study of these animals over a long period of time, carcinoma *in situ* may become invasive, metastasizing oral cancer.

Other workers attempted to produce changes in the lip mucosa of experimental rodents by repeated tobacco tar applications with and without the additional irritation of heat. When various tobacco products, such as whole tar or fractions of whole tar, were applied to the labial mucosa of mice every few days for many months, neoplastic changes did not occur at the application sites.[36] When acetone-solubilized tobacco tar was applied to the palatal mucosa of rats and mice at 56°C, hyperkeratinization and mild epithelial dysplasia occurred in both the experimental and control groups, but no neoplastic change was noted.[37]

Bastiaan and Reade[38] studied the changes in the lip mucosa of rats which had received repeated tobacco tar applications. The tar was prepared by using various kinds of standardized smoking machines. The experimental substances were applied by painting the mucosa of the lower lip between the two labial frena, placing a suture through the lips prior to the painting so that once the material was applied, all experimental substances could immediately be covered with a layer of Orabase® and the frena opposed by tying the suture. Animals were painted once every 7 days over a 280-day period or on alternate days for 56 days. In addition, in some groups of animals, heat at 60°C was applied to the mucosa either directly or through heating the experimental substance to the desired temperature. The most marked changes in the mucosa were seen when rats' lips were treated by heat plus the application of tar. The least amount of change occurred when only tar was applied to the oral mucosa. The tobacco tar extracts, which were repeatedly applied to the rat lip mucosa, produced a thickening in the oral mucosa resulting in an identifiable whitening of the affected area. In general, the more intense the irritation from tobacco tar and/or heat, the more marked the histologic changes. Tobacco tar had the ability to induce an irritational hyperplasia of the oral mucosa, but it did not have the capacity to initiate a neoplastic change. The application of heat to the experimental site further enhanced the hyperplastic response.

It has been shown that when tobacco tar is applied to the external skin of mice, it can initiate a neoplastic response. However, in most of the studies, tobacco tar that has been applied to the oral mucosa of mice and rats failed to initiate a neoplastic response. The question, of course, was whether this failure was due to some innate resistance of the oral mucosa of mice and rats to the mild carcinogens in tobacco or whether the saliva might have produced some kind of protection. As we shall see later, the saliva may play an important role in the carcinogenesis of the oral mucosa. In an interesting corollary study,[39] uncured raw areca nuts, active shell lime, and chewing tobacco, mixed in almost the same proportions as the product used for chewing, were made into a soft paste by grinding, and placed in the vagina daily. Some 40 animals survived from 324 to 380 days after the beginning of the experiment. Raised papillomatous growths were present in three animals, in which the histologic examination revealed malignant changes. In four animals there was thickening of the mucous membrane with definite malignant changes in the vaginal epithelium. Metastases were present in four animals. Apparently the vaginal mucosa is highly susceptible to the carcinogenic activity of "pan".

There is, thus, a considerable amount of work necessary to explain the epidemiological findings of malignant oral disease in man associated with tobacco chewing and smoking and the apparent resistance of the oral mucosa of animals exposed to the same carcinogens.

Chemical Carcinogenesis

Most of the research on chemical carcinogenesis utilized the polycyclic hydrocarbons in the skin of mice. As indicated earlier, the mouse became one of the animals of choice for the exploration of the pathogenesis of malignant change in epithelial cells. It was only natural, then, that when the field of oral oncology started to develop, the oral cancer researchers would apply the known skin carcinogens to the oral mucosa. If malignant disease developed, its pathogenesis could be studied. There are obvious similarities as well as differences between skin and oral mucosa. All epithelial surfaces or membranes are said to be protective, but the amount of protection they provide differs in different anatomical locations and is reflected in differences in structure. In some sites, the epithelial membranes perform additional functions, such as secretion and/or absorption. In these sites the structures are modified further. All epithelial coverings, however, do have features in common. They consist almost entirely of cells fitted closely together. Epithelial membranes contain no blood vessels, so the cells obtain nourishment from the tissue fluids of the underlying connective tissue. Almost all epithelial membranes possess a great capacity for multiplication, since almost all are subjected to some degree of "wear and tear" during their normal function, and they must therefore be capable of regeneration.

While various epithelial membranes have much in common, they also have differences. They differ in thickness, depending on whether the membrane that covers the surface is absorptive and/or whether there is much physical trauma. If, for example, there must be absorption or filtration, the membrane usually consists of a single layer of cells.

Superficial examination of the skin and the oral mucosa might lead to the belief that the two are identical structures, since both consist of a corium of fibrous connective tissue covered by a layer of stratified squamous epithelium. There are, however, distinct differences. The skin has appendages consisting of sebaceous glands, hair follicles, and sweat glands. Normal oral mucous membrane rarely has these structures. The appendages of the oral mucous membrane are mucous glands, salivary glands, and teeth. Other important differences are obvious when one examines the epithelium of wet and dry surfaces. The surface cells of wet stratified squamous epithelium probably remain alive, whereas the stratified epithelium covering of tissues exposed to air have nonliving cells on their outermost layers. The cells are not alive because they have become dehydrated. The surface cells of dry epithelium fuse together and become converted into the horny material, keratin.

The skin is the organ which separates man from his environment and allows him to adjust to environmental changes. It consists of two layers completely different in character and derived from different germ layers. The outer layer or epidermis is epithelium, and the inner or thicker layer, composed of dense connective tissue, is the dermis. The skin varies in thickness from less than 0.5 to 4.0 mm or more in different parts of the body. During embryonic development, cells of the developing epidermis invade the developing dermis to form sweat glands, hair follicles, sebaceous glands and, in some areas, nails. The wet epithelial linings of the digestive tube and of other internal passageways constitute, like the epidermis of the skin, a barrier between the community of cells that comprise the body and the outside world. One of the chief agencies ensuring the integrity of this wet epithelial membrane is its layer of mucus. From one end to the other, the digestive tube is richly provided with either individual cells or glands which produce mucus. Wet epithelial membranes thus equipped are termed "mucous membranes". The term "mucous membrane" includes not only the epithelial surface membrane, but also the underlying connective tissue which supports it, usually called the lamina propria. The lamina propria of mucous membranes lining the cheek, for instance, consists of a rather dense fibroelastic tissue that extends into the epithelium in the form of high papillae and merges deep into what is called the submucosa. Strands of this fibroelastic lamina propria penetrate through the elastic submucosa to join the fibroelastic tissue associated with the muscles of the cheek. There are numerous small mucous glands, some of which have a few secretory serous demilunes on the inner portion of the cheek. The oral mucous membrane also has its appendages, as previously mentioned.

When the epithelium of the skin of mice is painted with a carcinogen such as methylcholanthrene, carcinoma usually develops at the site of application. Cowdry[12] noted that painting the skin of newborn mice with methylcholanthrene produced no malignant change. However, when the skin of mice 5 days of age or older was painted with the same carcinogen, epidermoid carcinoma developed. There are marked differences between the skin of newborn mice and that of 5-day-old mice. The epidermis of newborn mice is thicker than that of older mice. The hair follicles and sebaceous glands of newborn mice are either absent or rudimentary. By the 5th day of postnatal life, however, sebaceous glands and hair follicles are well formed. Cowdry believed that a carcinogen needed a mode of entry in order for the epidermal cells to be affected and that the entry in the carcinogenesis of malignant disease of the skin was through the pits supported by the hair follicles. Lacassagne and Latarjet,[40] in attempting to verify the portal-of-entry theory, performed a simple experiment. They produced large but superficial wounds in the dorsal skin of mice. When these wounds healed, they healed without hair follicles or sebaceous glands. The skin was then painted with methylcholanthrene in such a way that the painting brush started on normal skin, crossed onto and across the scar, and continued onto normal skin. Epidermoid carcinomas developed on the normal skin, but the scar, which contained no appendages, was refractory to the carcinogenic action of the polycyclic hydrocarbon.

In an attempt to study the comparative reactions of skin and oral mucous membrane to the carcinogen 9,10-dimethyl-1,2-benzanthracene (DMBA), Levy and co-workers[41] applied the carcinogen with a single stroke to the skin side of the lower lip to the glabrous margin and with another single stroke to the mucous membrane side of the lip from the glabrous margin to the sulcus. Animals were sacrificed at various time intervals following the single application of the carcinogen, and the histologic changes on both the skin and mucous membrane side of the lower lip were studied. The comparative histologic study of the two surfaces following the application of the carcinogen revealed changes on the skin side consistent with those described by other investigators as characteristic of reactions to carcinogens. The mucosal surface side, however, remained entirely refractory to the application of DMBA and appeared normal. In postulating the reason for the resistance of the mucous membrane to carcinogens, they indicated that at least two factors had to be considered. The first was the lack of a portal of entry due to the absence of sebaceous glands in the mucous membrane. The second was the presence of a protective layer of mucus or saliva. At that time, they suggested that the use of antisialogogues and/or the extirpation of salivary glands in conjunction with the application of a carcinogen to the oral mucous membranes might help to clarify the problem. As mentioned earlier, there are considerable differences in the reactions of various animal species to the application and/or ingestion of carcinogenic chemicals, so another species or even a different strain of mouse might have responded differently.

Since the carcinogenic hydrocarbons exhibit a characteristic fluorescence in UV light, fluorescence microscopy was used to determine the distribution of the carcinogenic compounds in the skin and mucous membranes of hamsters.[42] The cheek pouch, mucosa, palate, tongue, and skin of the back of hamsters was painted three times a week for 15 weeks with DMBA in mineral oil. Following a single treatment, carcinogens detectable by their fluorescence were observed in the sebaceous glands. After two or three applications, the carcinogens were noted in heavy concentration in the epidermis. Examination of the mucous membranes, however, showed that there were fluorescent bodies only in the lamina propria beneath intact epithelial surfaces. Many of the fluorescent bodies appeared to be nucleated, but because the preparations were unstained and the frozen sections were thick, it was not possible to characterize the cells which contained the fluorescent bodies. Some appeared to be similar in size and shape to large histiocytes. As Salley pointed out, additional study was warranted in light of the fact that there was an absence of connective tissue neoplasms when hydrocarbon compounds were topically applied. With the exception of the skin, little or none of the fluorescent material was observed intraepithelially in that study.

It was not until 1965, almost 15 years after Levy and his co-workers suggested a possible protective role of saliva in oral carcinogenesis, that Wallenius investigated that possibility. Wallenius and his group did a rather exciting and comprehensive series of experiments aimed at investigating the role of saliva in oral carcinogenesis and at determining the necessity for using mice and/or hamsters as experimental animals to study oral malignant diseases. The rat had not been used as an experimental animal in the investigation of oral carcinogenesis; in fact, very little work had been done on skin carcinogenesis in this animal. Wallenius[43] used Sprague-Dawley rats. He painted their ears and oral mucosa three times a week with 0.5% DMBA in acetone. After 6 months of this treatment, squamous carcinomas of the ear appeared in 100% of the animals. In only one third of the animals treated for an average period of about 16 months did squamous carcinoma eventually appear in the oral cavity. It was assumed that the lower susceptibility of the oral mucosa to carcinogens was due to some protective intraoral mechanisms, and Wallenius proceeded to investigate this possibility.

In an effort to ascertain whether the low susceptibility of the oral mucosa of the rat to carcinogens was due to the anatomical character of the oral epithelium or to some external influences such as saliva, an ingenious experiment was devised in which the cheek skin of the rat was transplanted into its buccal cavity. This procedure enabled the researcher to paint the skin with all its appendages with a carcinogen in an environment of saliva. The intraoral skin was painted with DMBA three times weekly for 11 months. The same technique of painting was used on the contralateral intact buccal mucosa as well as on the external cheek skin. The intraoral painting with a carcinogen had no demonstrable effect on the transplanted cheek skin and produced only slight benign changes in the buccal mucosa. The external cheek skin, however, invariably reacted with the development of squamous carcinomas.

Wallenius concluded that the lack of sebaceous glands and hair follicles in the oral mucosa could not explain its low susceptibility to carcinogenic hydrocarbons. He then proceeded to test the hypothesis that the saliva protected the mucosa against the carcinogenic activity of polycyclic hydrocarbons. He reduced the secretion of saliva in rats by extirpation of the submaxillary and sublingual salivary glands and by bilateral ligation of Stensen's duct. The sialoadenectomy reduced the secretion of saliva by 90%. The remaining 10% was produced by the small mucous glands and appeared as a mucous layer on the surface of the oral mucosa. If, however, rats were given methylscopolamine, the oral mucosa was found to be dry for 8 hr.

During extensive experimentation, Wallenius found that the effect of methylscopolamine persisted unchanged for 1 year provided that consecutive injections were made at intervals of at least 48 hr. He then painted the mouths of rats with DMBA dissolved in acetone. In the animals with dry oral mucous membranes — that is, those in which there was both surgically reduced and pharmacologically inhibited secretion of saliva — squamous carcinomas invariably developed after 11 months of bilateral application of the DMBA. No carcinomas appeared in rats treated in the same way but with their salivary secretions intact. The incidence of malignant change was significantly lower in sialoadenectomized rats when compared with rats with dry oral mucosa either pharmacologically inhibited or surgically reduced and pharmacologically inhibited. In other words, there was some protection against chemical carcinogenesis in animals in which there was a reduction of saliva through the surgical removal of the major salivary glands due to the sparse mucus secretions from the small mucosal glands. Using fluorescence microscopy, he followed the penetration of rat oral mucosa by carcinogens in animals with intact or with inhibited salivary secretions. In rats with dry oral mucosa, but not in rats with normal salivary secretions, the carcinogenic hydrocarbon penetrated from the horny layer of the oral epithelium through the cellular layers to the lamina propria. The fluorescence was found to persist for at least 48 hr after application to the dry mucosa. He suggested that the protection by saliva against the carcinogenic hydrocarbon might be due to its rinsing or diluting effect but that the protection afforded by the sparse mucinous secretion originating from the small salivary glands in sialoadenectomized rats argued against that assumption. It was more likely that the mucinous content of the saliva acted as a protective barrier preventing the carcinogen from coming into contact with the underlying oral epithelium.[44]

In 1965, Fujino and co-workers[45] repeatedly painted the mucosa of the lower lip of mice with 4-nitroquinoline N-oxide (4NQO) in propylene glycol. Labial and/or lingual carcinoma occurred in 47% of the animals after 180 or more days of the experiment. When the application of 4NQO was accompanied by mechanical injury to the lower lip with a metal wire, 77% of the animals developed oral carcinoma in the same time period. Thus, another potent oral carcinogen was described: a water-soluble carcinogen.

Lekholm and Wallenius[46] wondered whether the penetrating power of chemical carcinogens into the oral mucosa was dependent on its lipid/water solubility coefficient. They tested the water soluble 4NQO by local applications in the same manner used to test the fat-soluble DMBA. Xerostomia was induced in rats by the use of methylscopolamine, and 4NQO was applied to the palate four times a week. The experimental model was designed to imitate the intermittent xerostomia which is established after the use of antihistamines, antihypertensive and diuretic drugs, barbiturates, tricyclic antidepressants and other psychopharmaceuticals. They found that rats treated with 4NQO developed palatal cancer after 7 months, whereas xerostomic rats developed palatal cancer after only 4.5 months, a reduction in time of tumor formation by 33%. Thus a protective effect of saliva exists whether the oral mucous membranes are attacked by water or by fat soluble carcinogens.

Other studies by Wallenius and co-workers[47] dealt with the effect of dietary zinc on the development of 4NQO-induced palatal cancer. They found that when rats were fed a zinc-supplemented diet and 4NQO was painted on the palatal mucosa, oral cancer occurred early. The animals had a shorter survival period than those fed a zinc-adequate or zinc-deficient diet and treated with 4NQO. Apparently a zinc-supplemented diet accelerated and a zinc-deficient diet retarded the development of chemically induced oral cancers in rats.

A well-developed model system has therefore become available for extensive studies of nutritional influences on oral malignancies. This model system might well be used to study metabolic imbalances, such as zinc deficiencies, as they affect malignant disease. It may also be useful in the study of pharmacologically active drugs as an influence on chemical carcinogenesis.

Carcinoma of the Tongue

Since the most aggressive oral cancers in man occur on the tongue, a model for lingual carcinoma seems desirable. One of the early studies on carcinoma of the tongue was reported by Levy,[48] who found that repeated painting of methylcholanthrene on the tongues of mice failed to produce carcinoma of the tongue, but that implantation of methylcholanthrene just beneath the surface epithelium of the tongue resulted in cancer formation. Even when a pure carcinogen such as DMBA was applied to the oral mucosa of the tongue, palate, and gingiva and held in position by cyanoacrylate, an effective mucous membrane adherent, the technique consistently produced carcinomas of the gingiva and palate of hamsters but not of the tongue.[49] An interesting finding was recorded by Carter et al. in 1967.[50] They described the unexpected induction of epithelial tumors of the tongue in mice following the application of 4-nitroquinoline 1-oxide (NQO) and croton oil to the dorsal skin of C57 × DBA2 strain mice. For 16 weeks, at weekly intervals, 0.2% NQO in acetone was applied to the dorsal skin. This was followed by weekly applications of croton oil for 53 weeks. Mice so treated were killed because of their poor condition or died at intervals ranging from 291 to 489 days after the beginning of the experiment. Only four skin tumors were produced by the treatment, but neoplasms of the tongue developed in 6 of 30 animals or in approximately 20%. Thus, an unexpectedly high proportion of C57 × DBA2 mice developed lingual carcinoma after the application of NQO and croton oil to the dorsal skin. The authors thought that this finding undoubtedly reflected a direct effect on the lingual epithelium of NQO acquired as a result of licking.

Dachi, after observing that DMBA penetrated through the cheek pouch epithelium of hamsters when it was dissolved in DMSO, applied 0.5% DMBA in DMSO to the tongues of hamsters three times a week for 30 weeks. All hamsters developed slow-growing plaquelike lesions of the tongue. However, the tongue lesions did not reach the size of other tumors of the oral mucosa which developed in adjacent areas. The mean latent period for the development of the tongue tumors was 94 days. Only six hamsters survived until the end of the 30-week experimental period because they died of oral mucosal tumors other than those that appeared on the tongue. Histologically, the tongues exhibited changes which developed from hyperplasia and dyskeratosis to carcinoma *in situ* and invasive carcinoma. However, frank carcinomas with invasion were seen in only 4 of the 15 hamsters studied. Larger, more anaplastic and more invasive carcinomas were seen on the lips, buccal mucosa, cheek pouches, and skin of the hamsters. There was either cellular atypia, carcinoma *in situ*, or invasive carcinoma in the esophagus of every animal.

Because of the long latent period required to induce lingual carcinomas, as well as the low incidence of such tumors, Fujita and co-workers[51.52] undertook to establish a method for the production of a high incidence of lingual cancer in a relatively short period of time. They made triweekly applications of 0.5% DMBA acetone solution to the left lateral border of the middle third of the tongue of 17 hamsters after scratching the area with a number 1 barbed broach (Group I). In a second group of 15 hamsters, the carcinogen was painted on the tongue without previous scarification. These two groups were adequately controlled by animals which were scratched and painted with acetone or painted with acetone without previous scratching. Their results were significant.

Group I animals showed a localized ulcer within 24 hr in the treated region. The ulcer persisted for a week to 10 days. Subsequently, the mucosa in the treated region became thick and rough. After 7 weeks of the experiment, a small nodular or papillary growth was noted at the treatment site in five of the hamsters. Similar lesions were found in 13 hamsters by the end of the 9th week; in all 17 hamsters, by the end of the 13th week. The lesions gradually increased in size, and all appeared grossly carcinomatous in 13 to 25 weeks. All hamsters died as a result of tumors by the end of the 27th week. Histologic examination revealed squamous carcinoma with extensive infiltration into the muscle tissue of the tongue. No metastasis was observed.

In the second group of animals (Group II), in which the DMBA was applied without previous scarification, ulceration was not observed, and the mucosa in the treated area gradually became rough, but to a much lesser degree than in the Group I animals. After 8 weeks, a nodular or papillary growth was noted in one animal. Similar lesions were observed in 13 hamsters by the end of the 11th week; and in all 15, by the end of the 13th week. As compared with the lesions in Group I, the lesions in Group II seemed to occur in multiple foci and were less localized. After 20 weeks, only one hamster had a grossly carcinomatous lesion. One animal died in 22 weeks, and three died by the end of the 27th week. Tumors continued to develop in the remaining animals rather slowly. Thus, all 17 hamsters in Group I developed lingual carcinomas by 27 weeks, but only 4 hamsters in Group II died as a result of tumors. The findings appeared to indicate that scratching hastens the formation of carcinoma induced by chemical carcinogens. Fujita and his group postulated that the ulceration or injury produced by scratching might have played an important role in the retention of the applied carcinogen in the area as well as in its penetration into the deeper layers of the mucosa. They also suggested that the regenerating mucosa of the ulcerated or injured area was more susceptible to the action of the chemical carcinogen. Since 100% of the animals scratched and painted with a carcinogen developed lingual cancer, they attempted to determine whether lingual carcinomas could be induced equally well in any site on the tongue or if there was a preferential disposition to specific sites, as in man.[51,52]

Hamsters were separated into five groups according to preselected sites of application: Group I, lateral border of middle third of the tongue; Group II, lateral border of anterior third of tongue; Group III, midportion of dorsum of tongue; Group IV, tip of tongue; and Group V, undersurface of tongue. The carcinogen was applied to each experimental site after scratching the area with a pulp canal reamer (Number 2 barbed broach) three times a week. The treatment was continued until carcinomas developed. The first neoplastic lesions noted were exclusively in the treated sites. The lingual carcinomas appeared to be responsible for the deaths of the hamsters. In some hamsters in Groups II, IV, and V, relatively large tumors developed in adjacent regions; however, none of these tumors reached the size of those in the various experimental sites on the tongue. The time period from the beginning of the experiment to the deaths of the animals caused by tumors was relatively short in Group I, much longer in Groups II, IV, and V, and very long in Group III, indicating that the lateral border of the middle third of the hamster tongue is the site most sensitive to carcinogenic action and is a predisposed site of lingual carcinomas in the hamster, as it is in man. In summary, lingual carcinomas in various sites of the hamster tongue resulted when the tongue was scratched at a particular experimental site with a pulp canal reamer and DMBA in acetone was applied three times a week. The period of carcinogenesis on the lateral border of the middle third of the tongue was relatively short; the period was longer in the lateral border of the anterior third and under surface and the tip of the tongue, and very long in the midportion of the dorsum of the tongue. Fujita and co-workers thus produced results which mimic the clinical manifestations of human lingual carcinomas and suggest that the use of this kind of experimental animal system could provide a useful model for the study of human lingual carcinoma.

Marefat and Shklar[53] undertook a series of studies to develop lingual carcinomas by the methods of Fujita et al., but introduced the use of DMBA in oil as a modification. They found that Fujita's technique of combining the application of DMBA in acetone with trauma was successful in producing carcinoma of the tongue in 100% of the hamsters by 12 to 15 weeks. The lesions became advanced by 15 to 16 weeks. The carcinomas were preceded by leukoplakic lesions comparable to those reported in hamster buccal pouch carcinogenesis and frequently noted in man. These researchers also found that it was important to use acetone as a solvent for DMBA, since DMBA in oil, together with the same kind of trauma used with DMBA in acetone, resulted only in epithelial dysplasia by 15 to 16 weeks. While they noted that the trauma accelerated the process of carcinogenesis, they also found that DMBA in acetone would produce lingual carcinoma by itself, although with a longer latent period. They pointed out that although carcinomas took an extra 2 to 4 weeks to develop when trauma did not precede the application of DMBA and acetone, the animals remained in better health, since they did not require the regular anesthesia necessary for lingual traumatization.

In further studies on the production of lingual cancer, Marefat and Shklar applied the technique of Fujita to two different strains of hamsters, namely, the Lakeview LSH and the Lakeview LVG strain.[54] A similar spectrum of lesions could be induced in both strains of animals but at different rates. It appeared that the rate of tumor development and growth was more rapid and the tumors were larger in the inbred animals. Leukoplakia lesions developed in both groups of animals at 4 to 6 weeks. Papillary growths were seen in two inbred animals (Lakeview LSH) at 6 to 7 weeks, but did not develop in the standard strain (Lakeview LVG) until 8 to 12 weeks. Further work by Marefat and Shklar[55] on the use of inbred strains of hamsters resulted in an excellent model, with leukoplakia and carcinoma developing as rapidly and as consistently in the tongue as in the buccal pouch model. The lingual carcinomas produced in the inbred strain of animals painted with DMBA and acetone were well differentiated. Attempts are currently underway to develop techniques for the production of less well differentiated invasive cancer of the tongue.

In further attempts to develop a model system for lingual carcinogenesis, Giunta and Shklar[56] applied DMBA powder covered and held in place with cyanoacrylate, DMBA in oil, and implants of DMBA to the lateral border of the tongues of Wistar rats. After 6 months of applications, there were no epidermoid carcinomas in any of the groups. Fibrosarcomas did develop in two of ten animals with DMBA implants.

Using an entirely different technique, Lijinsky and Taylor[57] discovered that the feeding of dichloro-nitrosopiperidine in the drinking water to 15 male rats resulted in the death of all animals before 24 weeks with tumors of the tongue, pharynx, esophagus, nonglandular stomach, nasal turbinates, and bronchial tree. When the dibromocompound was fed, all animals died of cancer by 41 weeks. When a threefold higher daily dose of the nitrosopiperidine was utilized, all animals were alive at 40 weeks, but all were dead by 55 weeks. Most of the animals died from tumors of the tongue, pharynx, esophagus, and nonglandular stomach.

Such studies warrant further consideration by the experimental oral oncologist, because the mode of action and the pathogenesis of tongue lesions under these circumstances have not been delineated. Lurie and Cutler,[58] using DMBA on hamster tongue, studied the effects of repeated low-dose-rate, high-dose-rate X-irradiation of the head and neck on lingual tumor induction by DMBA in outbred strains of the Syrian hamster. Animals received either topical application of 0.5% DMBA in acetone on the lateral middle third of the tongue three times a week for 15 consecutive weeks, 20R X-irradiation exposures of the head and neck once a week for 15 consecutive weeks, or concurrent radiation and DMBA treatments for 23 consecutive weeks. The researchers found that the tumor incidence was greater in the group treated with radiation plus DMBA (58%) than in the group treated with DMBA alone (38%). The mean tumor volumes in the groups treated with DMBA alone and with radiation plus DMBA did not differ significantly, probably because of the large standard deviation encountered in the group treated with radiation plus DMBA. Recalculating the mean log, tumor volumes to stabilize unequal variances resulted in a significant difference between the two groups, with tumors induced by DMBA alone being significantly larger. There was, however, no appreciable difference in the time of appearance of the earliest tumor between the group treated with DMBA alone and the group treated with DMBA plus radiation. The latent period was approximately 8 to 9 weeks in both groups. Nor were there differences in the progression of gross pathologic changes between the group treated only with DMBA and the group treated with DMBA plus radiation. Apparently there was some radiation-induced enhancement of lingual tumorigenesis, which was manifested by an increased number of papilloma-bearing animals. The exact nature of the mechanism to account for the radiation enhancement of DMBA-induced tumorigenesis has not been resolved and could provide a fertile area for further experimental research.

In another interesting study, Shklar and co-workers[59] investigated the effects of systemic administration of 13-*cis*-retinoic acid, a synthetic retinoid, on the DMBA production of lingual cancer. Other workers had demonstrated that vitamin A and/or retinoic acid were capable of inhibiting chemically induced tumors in the tracheal-bronchial tumor model and in a variety of experimental tumor systems. The findings of Shklar et al. demonstrated that animals receiving the analogue of vitamin A systemically exhibited a delay in the development of lingual tumors both grossly and microscopically. Frank carcinomas were noted 14 weeks following the application of DMBA in acetone to the lateral margin of the tongue of hamsters; however, in the DMBA-retinoid animals, only epithelial dysplasia and small areas of carcinoma *in situ* were found. Thus, another model has been developed which can be used by the experimental oncologist to study the interrelationship of nutrition and neoplasia.

Odontogenic and Other Jaw Tumors

In 1950, Levy and Ring[60] placed pellets of methylcholanthrene or dibenzanthracene in wax pellets at the surgically exposed growing end of the lower incisor teeth of hamsters and mice. The chemical carcinogens were prepared in wax pellets so that there was a slow release of the carcinogen. The animals were observed for periods of up to 1 year. The idea was to stimulate the odontogenic epithelium and/or the odontogenic mesenchyme in an effort to induce either ameloblastomas or other forms of odontogenic tumors. In hamsters, the tumors produced were either fibrosarcomas or rhabdomyosarcomas (Figures 1 to 4), whereas the tumors produced in mice were osteogenic sarcomas (Figure 5 to 9). The A strain mouse was used in these studies.

Ten years later, Green et al.[61] carefully placed methylcholanthrene in polyethylene glycol against the odontogenic soft tissues at the apical end of the lower incisor teeth of random-bred general purpose Bagg Swiss strain mice. Tumors developed between the 187th and 277th day postinjection. Histologically, the lesions were characterized by cyst formation and squamous metaplasia of the odontogenic epithelium. The squamous epithelium showed pleomorphism, hyperchromatism, atypical mitoses, and loss of polarity. The malignant epithelium proliferated as solid masses and strands into the surrounding muscle, bone, periodontal tissues, salivary glands, and nerves. In several instances, the cyst epithelium proliferated into the lumen of the cyst, giving a papillary configuration to the tumor. The researchers believed that the tumors were similar to those which spontaneously occurred in the Slye strain of mice described by Zegarelli as a malignant form of adamantoblastoma.[62] Some of the tumors produced in the animals of Green et al. showed a marked morphologic alteration in the form of spindle-shaped cells, which they thought were identical to the sarcoadamantoblastomas described by Zegarelli.

Mesrobian and Shklar[63] developed a technique for the study of carcinogenesis of the alveolar bone and the periapical and periodontal tissues in hamsters. They extracted the mandibular first molar and dipped the roots into pure DMBA powder; they

FIGURE 1.   Photomicrograph of a histologic section through the jaw of a hamster. A wax pellet containing DMBA had been placed at the apex of the growing incisor tooth 6 months earlier. The tumor was a fibro-sarcoma. (From Levy, B. M., and Ring, J. R., *Oral Surg. Oral Med. Oral Pathol.*, 3, 262, 1950. With permission.)

then replaced the tooth in its alveolar socket. Animals were sacrificed at 1, 2, 4, 8, and 12 weeks following the procedure, and the jaws were prepared for histologic examination. Tissues examined 1 and 2 weeks following DMBA implantation showed chronic inflammation. Starting at about 4 weeks after the DMBA implantation, malignant lymphomas developed. The lymphomas were ovoid masses of closely packed, deeply basophilic-staining cells located near the reinserted molars in the periosteal soft tissue at the inferior border of the mandible. Lymphoblastic cells at many levels of development, ranging from reticulum-type cells with large vesicular nuclei and prominent nucleoli to lymphocyte-type cells with dark nuclei and very little cytoplasm, were seen. A very reactive area of new bone formation suggestive of osteogenic sarcoma appeared between the lymphoma and the inferior border of the mandible. Within the mandible, malignant lymphoma cells were found in the marrow spaces adjacent to the implanted molar tooth. Two of the lymphoma-developing hamsters simultaneously developed invasive epidermoid carcinoma of the submandibular region. Each animal had a large prominent area of malignant lymphoma in what appeared to be the neoplastic remnant of preexisting lymph nodes. Therefore, the lymphogenous drainage of carcinogen from the experimental alveolar socket into the submandibular lymph node was postulated as being the probable site of initial action of the DMBA. The change within the lymph nodes was in addition to the less extensive intra-alveolar spread of the lymphoma. The investigators pointed out that the lymphoma might have started in the mandibular bone marrow and subsequently spread to the submandibular area. In any event, this rapid production of malignant lymphoma in all hamsters surviving 2 weeks or longer will facilitate experimental carcinogenesis. Subsequent study of this type of neoplasm might prove the value of such a system as a model for future research.

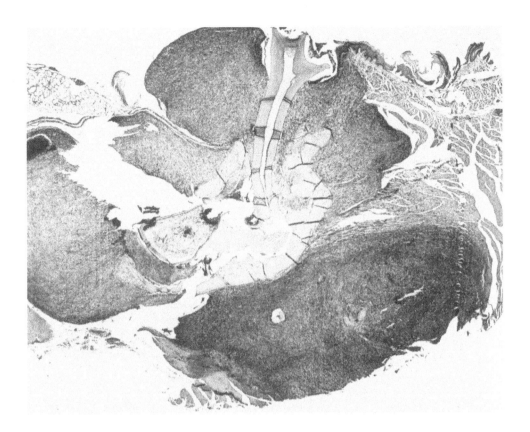

FIGURE 2. Photomicrograph of a section of the mandible of a hamster in which a pellet of methylcholanthrene in wax had been placed at the apex of the incisor tooth 6 months earlier. The tumor was a rhabdomyosarcoma. (From Levy, B. M., and Ring, J. R., *Oral Surg. Oral Med. Oral Pathol.*, 3, 262, 1950. With permission.)

Interested in producing osteogenic sarcoma, Sela and Harary[64] investigated the direct effect of methylcholanthrene on the tissues involved in the healing process of alveolar bone sockets in rats. Pellets of methylcholanthrene (MCA) in Vitepsol H-15 were introduced into the sockets of rats. Sutures were placed in order to prevent the loss of the carcinogen. Half of the rats in this group received three additional injections of emulsified pellets of 10% MCA in Vitepsol H-15 into the operation sites at 10, 20, and 30 days following the extraction of the molars. Other animals received a single injection of the emulsified MCA in Vitepsol H-15 into the operation site 20 days after the extraction. Crystalline methylcholanthrene was placed in the first molar extraction sockets of another group of animals. Some of these animals also received emulsified MCA in Vitepsol H-15 after 20 days, while others received three similar injections at 10, 20, and 30 days postextraction. All animals were maintained for 9 months.

The administration of MCA into the extraction sockets of rat mandibles induced fibrosarcomas. This appeared to surprise the investigators, since the work of Levy and Ring[60] indicated that osteogenic sarcomas could be induced in the mandible of mice using a somewhat similar technique. Sela and Harary failed to recognize that the technique was in essence quite different. Levy and Ring used an A strain mouse, while Sela and Harary used the Sabra strain albino rat. As indicated earlier, reaction to carcinogens, especially the polycyclic hydrocarbons, varies greatly from strain to strain within the same species and varies even further between species. The constant production of fibrosarcomas in all of the experiments reported by Sela and Harary highly

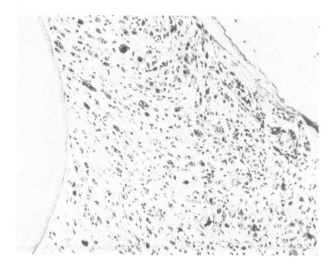

FIGURE 3.   Higher power of Figure 2 demonstrating the replacement of the periodontium by bizarre, poorly differentiated muscle cells. (From Levy, B. M., and Ring, J. R., *Oral Surg. Oral Med. Oral Pathol.*, 3, 262, 1950. With permission.)

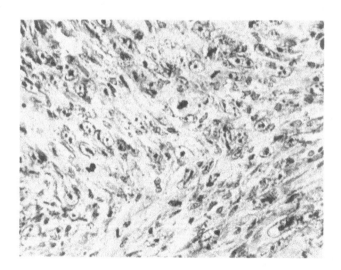

FIGURE 4.   Higher power of Figure 3 illustrating the abnormal nuclei, mitotic activity, and other features of a highly malignant, poorly differentiated rhabdomyosarcoma. (From Levy, B. M., and Ring, J. R., *Oral Surg. Oral Med. Oral Pathol.*, 3, 262, 1950. With permission.)

recommends their method for the production of fibrosarcomas in the jaws of rats, but the cause of the difficulty in inducing odontogenic tumors or osteogenic sarcomas remains undetermined.

In order to interpret the reaction of undifferentiated pulp tissue to the application of chemical carcinogens, it is necessary to have some understanding of the cell types involved. Cells are usually classified as intermitotic or postmitotic. The intermitotic cells are those which are capable of differentiating either into cells of a higher level or into vegetative cells which divide and then differentiate, such as basal epidermal cells. The differentiating intermitotic types are cells such as the spinous cells of the epidermis, which develop from cells of the vegetative type but can be distinguished from them. The postmitotic cells, on the other hand, discharge their function and then die (fixed postmitotic), or they discharge their function and revert to an intermitotic form (reverting postmitotic). The two types of intermitotic and the reverting postmitotic cells have potential malignant properties. Ameloblasts perform their function, revert to a squamous type of epithelium, and then die. These may be considered, then, as falling between the fixed and the reverting postmitotic types of cells; therefore, they have a limited potential for malignant change. More must be known concerning the biology of these cells before their resistance to carcinogenic agents such as DMBA can be explained.

Herrold[65] pointed out that the teeth of rodents are much more than masticatory organs; they are valuable biologic indicators, the development of which mirrors and records the metabolic status of animals. She investigated the hypothesis that the metabolic effect of such carcinogenic substances as N-methyl-N-nitrosourea (NMU) might be found in kymograph like records engraved in the teeth of rodents. Syrian hamsters 1 month old were used in the experiment. NMU solutions were freshly prepared prior to use. The chemical was dissolved in distilled water for intragastric administration and in physiologic saline for i.v. administration. Group A (16 animals) received 0.1 to to 0.2 mℓ (2.5 to 5 mg) of NMU i.v. into the lateral vein of the foot at monthly intervals for 3 to 4 months. Group B (10 animals) received NMU intragastrically (1 mg in 0.5 mℓ) twice a week for 4 months. Complete autopsies were performed on all animals killed or found dead, and the jaws were decalcified and prepared for histologic examination. The range of lifespan for Group A hamsters was 4 to 14 months with an average survival time of 8½ months. The survival time for Group B hamsters was 6 to 10 months, with an average of 8 months. Group A animals received 7.5 to 12.5 mg of NMU, while Group B animals received 32.0 mg. The jaws of all animals in both experimental groups revealed nests of cells, which varied in number and type in the periodontal membrane, predominantly of the lower incisor teeth. The most frequent type observed resembled the epithelial rests of Malassez. The peripheral cells in these oval nests showed a tendency to palisade. Herrold states that the epithelial rests had never been seen in the periodontal membrane of untreated hamsters before, but they had been noted in hamsters treated with diethylnitrosamine. Ameloblastomas developed in Group B animals.

With the exception of the ameloblastomas, which were found only in Group B animals, similar tumor types were observed in both experimental groups. Four animals had more than one type of tumor. Ameloblastomas were found in two of nine animals studied in Group B. Grossly, these animals showed unilateral enlargement of the mandibular region. The tumors had the characteristic microscopic appearance of an acanthomatous ameloblastoma, with interlacing strands or islands of cells imbedded in a connective tissue stroma. The cells lining the epithelial aggregates were odontogenic in character and were characterized by peripheral palisading. The central portion of the epithelial islands revealed keratinized and parakeratinized epithelial pearls. Cystic degeneration was frequent. The tumors were locally invasive and extended into the opposite side of the jaw, but they did not metastasize. In three other animals, nests of odontogenic epithelium located in the periodontal membrane of the mandibular incisors were noted. Histologically, these lesions bore a striking similarity to the ameloblastomas, and it is possible that they were an early stage in the development of the ameloblastoma.

Two of the surviving hamsters in Group A and three of nine in Group B developed ameloblastic fibromas, composed of ramifying islands and nests of epithelial cells in a loose, cellular connective tissue stroma. Throughout the tumor there were strands of cuboidal cells one or two cell layers thick which resembled the dental lamina. Ameloblastic odontomas, which showed ameloblastomalike islands of proliferating odontogenic epithelium, together with irregular zones of matrix resembling dentin and preenamel, were found in 3 of 14 animals in Group A and 4 of 9 in Group B. One of the animals in Group A had a complex odontoma, and one in Group B had a compound odontoma. Three animals (group not identified by the investigator) had cysts lined with noncornified squamous epithelium and an osteoblastic reaction of the periodontal membrane. In two animals, early cystic lesions developed in the stratum intermedium of the enamel organ. The cysts were lined with squamous epithelium, and the ameloblastic epithelium was atrophied.

In 11 of 14 Group A and 6 of 9 Group B animals, epidermoid carcinomas of the oral cavity, originating in the buccal mucosa, gingiva, tongue, and hard palate, were noted. Frequently the tumors were multicentric. All tumors were epidermoid carcinomas, but the degree of differentiation varied. Keratinization was prominent in some, whereas others were less well differentiated. The tumors were locally invasive and extended into the soft tissues of the face, alveolar bone, bone marrow, and periodontal membrane. Nests of tumor cells were noted in blood vessels, but in only one animal were metastatic nodules of epidermoid carcinoma found in the lung. In addition to the tumors, the epithelium of the oral cavity revealed focal areas of hyperkeratosis, basal cell hyperplasia, epithelial atypism, and squamous cell papillomas. A model system, capable of producing odontogenic tumors and intraoral carcinomas has been developed, in which studies of the effects of various metabolic imbalances and nutritional states on oral tumorigenesis can be studied.

Jasmin and co-workers[66] induced orofacial tumors in rats with radioactive cerium chloride. They injected 59 6-week-old Sprague-Dawley male rats near the angle of the mandibular bone with radioactive colloidal cerium. Of the 59 animals injected, 42 exhibited single tumors, and 11 had two tumors. (The other six animals either died spontaneously of bronchopneumonia or were killed because they were moribund with bronchopneumonia.) Of the tumors exhibited by the animals, squamous cell carcinoma was the most frequent type (42 in number), and soft tissue and bone sarcomas were the second most frequent type noted (14 in number). There were three salivary adenocarcinomas, three tumors with "double differentiation", one odontosarcoma, and one ganglioneuroblastoma. All the squamous carcinomas were well differentiated. Most were located at the left posterior intermaxillary corner near the site of injection. Six tumors originated at the inferior retromolar region, three at the superior retromolar region, and three in the jaw. These tumors apparently invaded the floor of the mouth and/or the sub- and retromaxillary regions and invaded nerves. The soft tissue sarcomas, which were mostly fibrosarcomas, also appeared to be located in or to invade the floor of the mouth. There were four osteogenic osteosarcomas, one of which metastasized, and a single odontosarcoma, which resembled human odontosarcoma, with ameloblastic proliferation and elaboration of dental mineralized tissue. Thus, a wide variety of tumors was produced in the jaws and oral cavity of rats injected with radioactive cerium chloride and maintained from 262 to 501 days.

Viral Oncology

In 1910, Rous,[67] working at the Rockefeller Institute, isolated a virus from a chicken sarcoma which, when injected into other chickens, produced a sarcoma at the site of injection. This was one of the first viruses associated with the transmission of a malignant disease and opened the era of viral oncology.

The story of the virus etiology of malignant disease is an interesting and complex one. Shope[68] of the Rockefeller Institute, who was a great hunter, shot a rabbit with tumors under the skin of the front and hind legs. He was able to transmit the tumor by the use of a cell-free extract, indicating a viral etiology. The tumor did not kill the animal and eventually regressed. The virus was later shown to be mosquito-transmitted and related to the myxoma virus. Later, Kilham[69] showed that he could transform a rabbit fibroma virus into a myxoma virus by adding heated myxoma virus to live fibroma virus. The resulting lesion was a myxofibroma, a viral-induced tumor of rabbits still used in the study of viral oncogenesis. He also showed[70] that a fibroma of the grey squirrel could be transmitted by a cell-free extract and that this virus-induced tumor also had a mosquito vector.

Wild cottontail rabbits in Kansas and Iowa frequently develop tumors of the skin, especially around the neck and shoulders. These tumors appear as large horny warts. Shope[71] found that these horny warts could be transmitted by rubbing cell-free filtrates into the scarified skin of rabbits. The tumors, however, could not be produced in the mouth or in other organs. Parsons and Kidd,[72] also working at the Rockefeller Institute, showed that a benign papilloma, which occasionally arose under the tongue of rabbits, was viral induced. The virus was extremely specific, since it reacted only on the under surface of the tongue of rabbits. It was quite different from the Shope papilloma virus because animals immune to one were susceptible to the other.

Oral papillomas of dogs were also induced by a virus.[73] The virus was specific for dogs. If it was injected into the mouths of kittens, mice, rats, guinea pigs, or monkeys, no tumors resulted. Warts in cattle and cutaneous papillomas of horses were also shown to be viral induced.[68] Lucke,[74] worked on carcinoma of the kidney in leopard frogs and found that it was a virus-induced tumor. Other virus-induced tumors are the venereal sarcoma of the dog and the fox and a remarkable number of fowl leukemias and solid tumors.

A virus isolated from a parotid tumor of mice began another surge of excitement and activity in viral oncology.[75] When this virus was injected into newborn mice or hamsters, tumors developed in all organs. The virus was therefore named "polyoma virus", and a considerable amount of work with this virus has been done since 1951. The use of newborn animals for the study of viral oncology was established, and many viruses isolated from humans and monkeys were shown to be oncogenic in newborn hamsters and mice.

The evidence for human oncogenic viruses is slowly accumulating. There are many examples of malignant solid tumors and leukemia in families in successive generations. The accidental transmission of malignant disease from man to man has also been reported. Viruses isolated from jaw tumors of patients with Burkitt's lymphoma and from some patients with nasopharyngeal carcinoma (Epstein-Barr virus) have been shown to produce lymphomas in certain strains of marmosets. Recently Anthony Jahn, reporting for the Oncology Times (October 1980) on the Symposium on Viruses and Cancer held during the 1980 International Symposium on Cancer, pointed out that although the association of viruses and human cancer was clear, host factors were the major determinant as to whether the oncogenic potential of a virus was fulfilled. "Transformation-associated" cellular proteins isolated from chemically and/or virally induced tumors were shown to be closely related, if not identical. Many oncogenic viruses were reported to exert immunosuppressive effects on the host, thus impairing their normal immunosurveillance mechanisms and allowing the development of malignancies.

In summary, practically all fowl tumors, most mouse tumors including leukemias, and the majority of rat tumors and leukemias can be transmitted by sterile filtrates or a recognized virus. Many rabbit, dog, squirrel, cattle, frog, and fish tumors have also

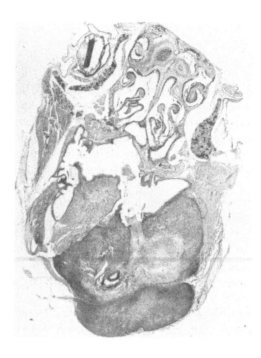

FIGURE 5.   Photomicrograph of a sagittal section
through the head and tumor of a mouse. A wax pellet
containing methylcholanthrene had been inserted into
the apical area of the lower incisor tooth 6 months
earlier. The tumor was an osteosarcoma. (From Levy,
B. M., and Ring, J. R., *Oral Surg. Oral Med. Oral
Pathol.*, 3, 262, 1950. With permission).

been transmitted by sterile filtrates. There is increasing evidence that some human
tumors may be transmissible by a virus. It was natural, then, that experimental oral
oncologists would study the oncogenic potential of viruses in oral neoplasms.

Viral Oral Oncology

As early as 1964, Stanley and co-workers[76] injected polyoma virus into mice and
reported on some 70 tumors present in 11 mouse heads in one study and a large number
of tumors in a different strain of mouse using a substrain of the virus in another
study.[77] Lucas[78] injected newborn C_3H strain mice subcutaneously with the L_1D/I
strain polyoma virus. Animals were killed at varying intervals up to 1 year, and the
tumors were carefully studied. (Precise numbers of animals used were not given in the
publication.) Tumors occurred in close connection with the incisor and molar teeth of
both the mandible and the maxilla. In some cases, there were multiple tumors. In
general, they resembled those described by Stanley et al. Almost all of the tumors
appeared to have a connection with the gingival epithelium. Some of the tumors which
appeared to have a connection with the incisor teeth gave an appearance somewhat
reminiscent of the follicles of human ameloblastoma, with central areas of stellate
reticulumlike cells. Lucas pointed out that any possible analogy between the human
ameloblastoma and this tumor should not be pursued too far, since in the mouse tumor
the cells were not really stellate in type, and in fact the appearances could be due to
intercellular edema. The other feature comparable to ameloblastomas was cyst formation. Small cysts were quite frequently seen, but larger ones surrounded by a thin
layer of tumor tissue also developed. Tumors of that type seemed to have quite a

159

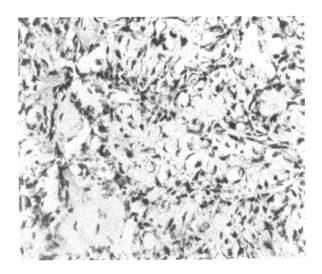

FIGURE 6. High power of the tumor seen in Figure 5, showing
the abnormal osteoid produced by the tumor cells. (From Levy,
B. M., and Ring, J. R., *Oral Surg. Oral Med. Oral Pathol.*, 3,
262, 1950. With permission.)

noticeable resemblance to the monocystic type of ameloblastoma in man, with areas
of squamous change and keratin production. Even in these tumors and cysts, a connection could be found between the tumor and the gingival epithelium. Stanley and
his group considered that the tumors arose from gingival epithelium, though often the
growths were so extensive by the time of autopsy that the actual site of origin could
not be detected. Lucas' study indicated that the tumors do arise from gingival epithelium, although the possibility of an origin from epithelial rests was also considered.
Main and Dawe,[79] on the basis of ingenious and carefully executed transplantation
experiments, believed that the tumors arose from the outer enamel epithelium. Under
sterile conditions, 14-day mouse embryos were removed from the uterus, and the maxillary and mandibular incisor tooth buds were dissected out. These were then placed
in virus suspension and subsequently transplanted s.c. into newborn syngeneic mice.
Tumors appeared in the transplants after 20 days and were identical to those described
by Stanley et al.[76] Early tumors could be seen arising from the outer enamel epithelium.
Interestingly enough, tumors also appeared in the host mice, arising in the outer enamel epithelium. In the tumors described by Lucas, the origin, where it could be detected, appeared to be the fully differentiated squamous epithelium of the gingiva.
There is still considerable work to be done concerning the role of the polyoma virus
in odontogenic tumorigenesis in various strains of mice.

In an attempt to develop a primate model for oral oncogenic studies, Levy et al.[80]
made an intraosseous inoculation of the Rous sarcoma virus into the mandibles of five
marmosets (three *Callithrix jacchus*, two *Saguinus oedipus* and one *S. fuscicollis*) at 1
to 9 days of age. In two of the *C. jacchus* animals (which survived 78 and 131 days),
mandibular tumors developed at the site of inoculation (Figure 7). All were biopsied
and found to be typical osteosarcomas. They were composed of anaplastic fibroblasts,
masses of abnormal osteoid, and spicules of poorly formed bone. Because the tumors
interfered with eating, the animals were killed. No metastatic lesions were found at
autopsy. X-ray examination of the jaw revealed a radio-opaque mass with destruction
of erupted and unerupted teeth. There was a typical "sunburst" radiating in all directions from the central densely sclerotic mass (Figure 8). Histologically the tumor was

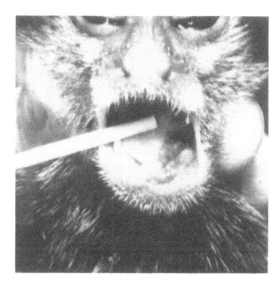

FIGURE 7.   Gross appearance of an osteosarcoma resulting from the injection of Rous sarcoma virus into the mandible of a *Callithrix jacchus.*

composed of plump anaplastic fibroblasts, osteoblasts, and multinucleated giant cells
with large amounts of neoplastic osteoid and poorly formed bone (Figure 9). The tumor eroded the erupted and unerupted teeth, leaving small islands of dentin. In some
areas, the unerupted permanent tooth was completely destroyed. Thus, osteogenic sarcoma of the jaws of very young primates was produced, providing another model
system for the study of malignant jaw lesions and the way in which they erode, invade,
and destroy tooth and bone substance.

Gingival Cancer

Epidemiological studies have indicated that most investigators do not distinguish
between cancer of the gingiva and cancer of the alveolar ridge despite the fact that the
mucosa covering the gingiva and the dentulous and/or edentulous alveolar ridge are
different in their histological structure. All cancers in this general area are usually
grouped under either cancer of the gingiva or cancer of the "gums". In one of the
largest samples of gingival (alveolar ridge) carcinomas studied,[81] 77% of some 600
patients were men and 23% women; 79% of the women and 88% of the men were
between 50 and 80 years of age. On the other hand, Srivastava and Sharma[82] studied
85 Indian patients with cancer of the gums and reported that 72% were younger than
50 years old.

Most of the epidemiologic studies showed that about three quarters of the gingival
cancers occurred on the mandibular gingiva, whereas some 20 to 25% occurred in the
maxilla. Adjacent structures were frequently involved. In the mandible, the tumor usually spread to the mandibular sulcus and the floor of the mouth. In the maxilla, the
tumor frequently involved the palate, maxillary sulcus, and/or buccal mucosa. The
use of tobacco was implicated as an etiologic factor in carcinoma of the gingiva. Despite the fact that carcinoma of the gingiva is a relatively common oral cancer with a
high mortality rate, little experimental research has been done with this tumor.

Experimental Gingival Carcinogenesis

Zussman[83] noted that an infiltrating gingival carcinoma was consistently found in
rats that had ingested 0.03% *N*-2-fluorenylacetamide (2FAA) added to a low pyridox-

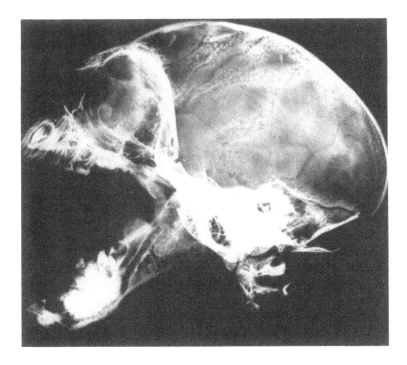

FIGURE 8.   Radiographic picture of the osteosarcoma of the mandible of *Callithrix jacchus*. (From Levy, B. M., Taylor, A. C., Hampton, S., and Thoma, G. W., *Cancer Res.*, 29, 2237, 1969. With permission.)

ine diet and later stimulated with stilbestrol. After the rats (W/Fu strain) had ingested 2FAA in their diet for 3 months, 5 or 10 mg pellets of stilbestrol were implanted s.c. and changed monthly. The tumors induced in this study were adenocarcinomas. In another series of experiments, Zussman[84] fed W/Fu rats 0.03% 2FAA starting at 1 and 3 months of age. The 2FAA was added to a low pyridoxine diet in some animals and to a normal laboratory chow in others. Animals received s.c. implants of 5 or 10 mg pellets of diethylstilbestrol 1 month later, while still on the 2FAA low pyridoxine or normal diet. Of 24 W/Fu rats fed 2FAA, 17 survived the 5 or 11 months during which the diethylstilbestrol was admininstered. He found that five of eight animals on the 2FAA low pyridoxine 5 mg diethylstilbestrol regimen had infiltrating gingival carcinoma by 12 months of age, that seven of eight 2FAA, low pyridoxine, 10 mg diethylstilbestrol had infiltrating gingival carcinoma at 9 months. Degenerative changes were also seen in the sebaceous and salivary glands examined. Extraoral tumors were also described. The important point, however, is that following this regimen, diffusely infiltrating gingival carcinomas developed. Many areas of tumor formation histologically resembled cylindromatous carcinoma. The invasive nature and location were consistent with primary tumors of the minor salivary glands. The uniform appearance and cystic arrangement of the cells were noted where tumor invaded bone, teeth, and gingiva. Adjacent tumor areas were composed of squamous neoplasms. It was believed that a salivary gland origin could account for only portions of the tumor.

The enamel organs of the animals in Zussman's investigation were examined carefully for odontogenic tumors. Distortion of the outermost stellate reticulum cells, with loss of polarity and increased nuclear size and invasion of the surrounding connective tissue and periosteal tissues by blood vessels, was noted. There appeared to be a histologic resemblance of these changes to the squamous areas of the tumor. In advanced lesions, the enamel organs appeared to be replaced by tumor. Since the tumor was

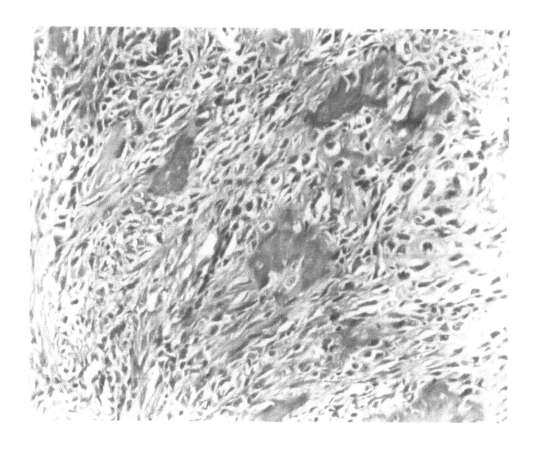

FIGURE 9.   Photomicrograph demonstrating the histologic features of the osteosarcoma induced in the mandible of the marmoset, *Callithrix jacchus,* by the Rous sarcoma virus. (From Levy, B. M., Taylor, A. C., Hampton, S., and Thoma, G. W., *Cancer Res.,* 29, 2237, 1969. With permission.)

present in both the upper and lower jaws and infiltrated bone, teeth, and gingiva, the odontogenic epithelium and minor salivary glands were considered by Zussman to be the sources of origin of the malignant neoplasms. Just how 2FAA in the presence of a low pyridoxine diet produced these unusual lesions has not been elucidated. Neither has the relationship of diethylstilbestrol to the occurrence of the tumor been thoroughly examined. Both are fruitful areas of future research.

Shklar and his group also studied gingival carcinogenesis. In a study reported by Al-Ani and Shklar,[85] 30 Syrian hamsters 3 to 4 months of age were utilized. The gingival tissues around the mandibular molar teeth were painted three times weekly with 0.5% DMBA in heavy mineral oil. The tooth surfaces were painted so that some of the carcinogen entered the gingival sulcus. Animals were sacrificed after 4 months, and autopsies were performed. No tumors were seen on the control (unpainted) side of the animal. On the experimental side, however, the gingival surface appeared rough and pebbly in localized areas, and there were obvious tumor masses in 11 of the animals examined. Histologically, the interproximal gingiva was markedly hyperplastic and hyperkeratotic. In the area of the epithelial attachment, the tissue was hyperplastic so that the attachment was much broader than normal. The epithelial attachment appeared to be on the cementum instead of the cementoenamel junction. On the mesial gingiva of the experimental side, both benign and malignant epithelial neoplasms developed. Invariably a hyperkeratosis with some epithelial proliferation, characterized by downward extension of the rete pegs, appeared in all animals. In three animals, there were large well differentiated squamous cell carcinomas of the marginal and attached gingiva. The sulcus epithelium, however, was relatively unaffected in most animals. Only a few animals presented with carcinomas after 4 months of continuous painting with the DMBA in oil. However, there were numerous areas of dyskeratosis and papillomatosis during the 4-month experimental period.

Later Mesrobian and Shklar[86] attempted to speed up the induction and increase the yield of epidermoid carcinomas by prolonging the retention of each application of DMBA on the gingival surface through the use of cyanoacrylate tissue adhesives. They used 50 male and female Syrian hamsters 3 to 4 months of age. Of the group, 20 hamsters (Group I) were treated with weekly gingival applications of pure DMBA powder. The carcinogen was placed on the mandibular labial gingiva in the incisor area and held in place by a drop of cyanoacrylate liquid; 10 animals (Group II) received weekly application of DMBA without cyanoacrylate adhesive. Other controls received weekly cyanoacrylate dressing without carcinogen (Group III), and some were untreated (Group IV). None of the control animals throughout the experimental period showed change referable to neoplastic development. The experiment was continued through 20 weeks, at which time the animals were sacrificed for gross and microscopic examination. Starting at the 5th week of the experiment, all the experimental hamsters and those controls receiving cyanoacrylate adhesive showed moderate gingival reddening. However, by the 10th week, while there was gingival reddening and some bleeding in Group I experimental hamsters, all of the controls were normal. At 10 weeks, 20% of Group I hamsters had easily discernible tumors in the incisal area of the gingiva. From the 11th to 20th week, 60% of the experimental hamsters had whitening and thickening of the epithelium, and 30% had obvious gross tumors of the gingiva. All controls were essentially normal. At 16 to 20 weeks, biopsies of the experimental hamsters' gingiva showed frank invasive epidermoid carcinoma. Some of the carcinomas were hyperkeratotic and exophytic, while others were less differentiated and infiltrative. Autopsies of the experimental hamsters showed no gross or microscopic evidence of distant tumor metastases.

Thus Shklar and co-workers pointed out that the induction time for the development of mucosal cancer in hamsters could be shortened by using cyanoacrylate adhesives. This would facilitate the production and study of many types of oral surface malignancies, especially those that were produced with difficulty by earlier researchers. They postulated that previous difficulties were the result of the carcinogens being washed away from the mucosal surface by saliva. The continued presence of the carcinogen on the mucosal surface resulted in epidermoid carcinoma. Thus, the concept of a necessary portal of entry such as mucosal ulceration or hair follicles is not applicable to the production of malignant disease of the gingiva. Results by others who induced malignant change in the oral mucosa of mice only after desalivation and subsequent ulceration were due to the absence of normal salivation, which prevented the carcinogen from being washed away.

There are, of course, many questions that are left unanswered in these experiments. While it may be reasonable to postulate that a portal of entry is not necessary if the carcinogen is held in place by an adhesive such as the cyanoacrylates, it is also important to point out that the saliva may do more than simply "wash away" the chemical carcinogen. The work of Wallenius and others indicated that saliva or the mucus of the saliva formed a protective "coating" which could be destroyed by the cyanoacrylate. Simply wiping the gingiva free of saliva prior to the application of DMBA and cyanoacrylate might assure the loss of the salivary protective factor so that carcinogenesis occurred. There is obviously considerable work that can be done on the relationship of saliva, the salivary glands, and oral mucous membrane carcinogenesis.

SALIVARY GLANDS

Tumors of the major salivary glands account for 5% of all benign and malignant tumors, excluding those of the skin. Tumors of the minor salivary glands are much less common than those of the major glands, accounting for some 20% of all salivary gland tumors.[87] It has been suggested that the relative frequency of tumors of the minor salivary glands may be even greater than that in some areas such as South Africa, Uganda, the West Indies, and India.[88] Of the three major salivary glands, the parotid is much more often the site of tumors. Between 85 and 90% of all tumors of the salivary glands occur in the parotid. Most of the remaining tumors occur in the submandibular gland, whereas tumors of the sublingual gland are relatively rare. The glands of the palate are more frequently the site of tumors than any of the other intraoral glands, accounting for about 55% of all intraoral salivary gland tumors. About 15% of the intraoral salivary gland tumors occur on the lip.

As part of an across-the-board survey of the incidence of second cancers, Berg et al.[89] found a close, unexpected, and highly significant association between breast cancer and salivary gland cancer. Women who had a salivary gland carcinoma had an eightfold risk of having a subsequent breast carcinoma develop. Women with anaplastic epidermoid carcinomas of the salivary gland had a 13.5-fold risk of having breast cancer while women with other types of salivary gland cancer had a fourfold risk. They noted one other curious finding. In every instance of the mucoepidermoid subset, the breast cancer occurred on the same side as the salivary gland cancer. If this were a chance phenomenon, it would occur only once in about every 30 sets of observations on laterality. Because of the high incidence of malignant disease of the salivary glands and this unique association of the salivary gland cancer with breast cancer, studies of the pathogenesis of salivary gland cancers are of extreme importance in the biology of malignant disease. One additional study might serve to point up the importance of the experimental salivary gland tumor model to cancer biology. Brown and Frankel[90] chemically induced tumors in the submaxillary glands of rats. Since it had been reported that the saliva of humans with oral cancer contained markedly elevated levels of secretory antibody, they tested the rats' saliva for blocking and/or potentiating "antibody" activity. The results of their studies indicated that saliva from tumor-bearing animals potentiated tumor cell killing of either tumor cells or normal rat leukocytes when pretreated with the tumor-bearing animals' saliva. The effect was much more marked when the tumor cells were pretreated. Control saliva occasionally produced a similar effect, but not as consistently nor with the potency of the saliva from tumor-bearing animals. They inferred a new system of immunological surveillance, the arming or potentiation of leukocytes by soluble tumor antigens in glandular secretions.

Salivary Gland Carcinogenesis

Much of the early work in experimental salivary gland carcinogenesis was done by producing salivary gland tumors by feeding various compounds to mice and/or rats. In 1941 Wilson and his associates[91] fed 2-acetylaminofluorene to rats, which developed tumors in many organs, but none in the salivary glands. Five years later, Heiman and Meisel[92] gave 59 2½ to 3-month-old Wistar rats 10 mg 2-acetylaminofluorene in peanut oil by gavage every other day. Later, 20 mg were given every 3rd day. The total amount ranged between 250 and 650 mg over 111 to 227 days. Of the treated rats, 22 developed nodular swellings in one or both of the submaxillary regions. Autopsy revealed enlarged submaxillary glands and many small discrete or large confluent cysts containing a yellowish oily fluid (peanut oil). On dissection, the swellings were traced to the mediastinum. In addition to the cystic swellings, two adenocarcinomas and three adenomas of the submaxillary gland were found. The adenocarcinomas had distorted architecture, which included large cysts containing papillary projections of epithelium. Many of the cystic spaces were completely filled with masses of epithelium. There was also invasion of the s.c. connective tissue by groups of cancer cells which were irregularly cuboidal. The cytoplasm of these cells contained mucus. The adenomas of the submaxillary gland appeared as isolated areas of hyperplastic ducts and acini, in which the normal cuboidal epithelium was replaced by columnar epithelium. The localized adenomas showed a transition from the normal tissue to the adenocarcinoma described above.

Some 20 years later, Zussman[93] fed W/Fu rats N-2-fluorenylacetamide (2FAA) in a low pyridoxine diet. One month later, 20 male and four female rats were treated with s.c. transplants of 5 or 10 mg pellets of diethylstilbestrol while still on the specified diet. The pellets were changed monthly. As described above, rats ingesting 2FAA in a low pyridoxine diet and later stimulated with diethylstilbestrol had diffusely infiltrating gingival carcinomas.[84] Zussman pointed out that in order to understand the gingival lesions, it was important to study alterations in the minor salivary glands and in the sebaceous glands surrounding the middle ear of the rats. During the earlier study, he had found that there were variations in the frequency and site of tumor formation in different strains of rats. In the report under consideration here, the salivary and sebaceous glands were carefully studied in W/Fu rats. He found that although the minor salivary glands are extensively distributed in the oral cavity of the rats, the only overall alteration noted when the experimental protocol was applied was atrophy, which seemed more marked than that found in the major salivary glands. Atrophy and carcinomas occured in the parotid, submaxillary, and sublingual salivary glands. Only cystic atrophy arose in the sebaceous glands that line and surround the middle ears.

Carcinosarcomas of the submaxillary salivary glands in castrated female AXC rats receiving N,N-fluorenyldiacetamide (2-Fdi AA) and norethandrolone were reported by Reuber.[94] Carcinosarcomas of the salivary gland did not develop in intact animals of either sex and were therefore thought to be related to the anabolic effects of norethandrolone. Salivary gland tumors did not occur in similarly treated castrated male rats with or without norethandrolone, nor did tumors develop in intact females given norethandrolone or in intact or castrated female rats given progesterone or testosterone. Apparently both the absence of the gonads and the presence of norethandrolone were of importance in the induction of submaxillary salivary gland carcinomas in this strain of rat when fed 2-Fdi AA. The tumors were quite large and grossly appeared to replace most of the gland. Dilatation and distortion of the ducts and atrophy or disappearance of the acini were noted in some portions of the gland, while in others tumor cells appeared to be arising from the duct walls and invading the adjacent stroma. Histologically, the tumors were composed of two types of cells, epithelial and stromal. Much of each tumor was composed of spindle-shaped cells spreading in several directions. The nuclei were large, vesicular and hyperchromatic, oval or round with one or two prominent nucleoli. In a few areas, myxoid tissue or foci of hyalin matrix were noted. Scattered tumor giant cells with two or three nuclei and many mitotic figures, often atypical, were seen. No cross striations were noted. The epithelial component was composed of polygonal cells with dark eosinophilic or at times basophilic cytoplasm and large hyperchromatic oval nuclei. These epithelial cells were frequently arranged in sheets or islands, but in some areas they were interspersed with the spindle-shaped or undifferentiated type of cell. Around the edges and within the periphery of the tumor were small acini and ducts.

The alkylating agent methyl nitrosourea (MNU), which is a potent bacterial mitogen, induced tumors in a wide variety of organs when injected as a single intravenous dose into rats. However, no MNU-induced salivary gland tumors were reported. Parkin and Neale[95] recognized that since the half-life of MNU after i.v. injection into a rat was only 4 min, the period of exposure to the carcinogen could be defined with considerable precision. They pretreated rats with isoprenaline sulfate (IPR) and then injected the rats i.v. with MNU at times chosen to coincide with different phases of the IPR-induced cycle of metabolic events. A single i.v. dose of the unstable carcinogen MNU, given at 28 or 31 hr after IPR treatment, gave rise to the first salivary gland tumors to be observed following MNU administration to rats. No salivary gland tumors were observed in rats given MNU 8 hr after IPR treatment or in rats treated with either MNU or IPR alone. The relevance of the IPR pretreatment resides in the synchronized burst of DNA synthesis produced in the salivary glands. By the time of maximum DNA synthesis, other IPR-induced metabolic changes, including increased levels of cAMP and membrane-bound adenylate cyclase and the stimulation of protein, RNA, and glycogen synthesis, had reverted to normal. In view of the high level of DNA replication observed in the salivary glands 28 or 31 hr after IPR treatment, the fact that only 13% of the rats surviving 10 weeks or more produced tumors was disappointing to the researchers. They pointed out that 28 to 100% of rats treated with DMBA produced salivary gland tumors. (These experiments will be discussed later.) However, they also explained that the half-life of MNU was a matter of minutes, whereas the exposure to DMBA was prolonged, since DMBA persisted at the site of injection for some weeks. Further, a sex difference was observed in the induction of carcinomas by DMBA, indicating that males were much more susceptible to the action of the carcinogen than were females. In the IPR-MNU experiments, only females were used. It is likely that the short carcinogen exposure and the use of female rats may have resulted in the low yield of tumors.

Shortly after the polycyclic hydrocarbons were made available to the scientific community (1940 to 1960), a rash of papers appeared in which both carcinomas and sarcomas were reported to have been induced in the salivary glands of mice, rats, and guinea pigs by the local application of the various carcinogens.[96-101] Carcinomas were induced more frequently than sarcomas and generally took a shorter time to develop. Most of the carcinomas were of the squamous cell type, although adenoacanthomas and an occasional adenocarcinoma were reported in mice. The usual method of application of the carcinogen was by means of a pellet of cholesterol or wax which contained the polycyclic hydrocarbon.

Glucksman and Cherry,[102] in a study of the effects of ionizing radiation on the salivary gland, reported a sex difference. This observation led them to investigate the effects of endocrines on the carcinogenesis induced in salivary glands by the polycyclic hydrocarbons, particularly in male rats.[103] A group of 129 male and female black-hooded rats, 2 to 5 months old, were used for the experiments. In the experiments, 0.1 mℓ of a saturated solution of DMBA in acetone or a 1% solution of the carcinogen in olive oil was injected into the salivary gland complex on one or both sides of the neck with or without surgical exposure of the salivary glands. An attempt was made to deposit the carcinogen in the submandibular, sublingual, and parotid glands. Animals were killed at 1, 3, 5, 7, 10, and 14 days after injection, then at weekly intervals for 5 weeks, and thereafter at periods varying from 2 to 8 months, whenever tumors or other conditions became insufferable. The rats treated with DMBA in olive oil were killed at intervals ranging from 54 to 477 days after injection, when tumors likely to cause death or suffering developed.

Injections of DMBA in acetone caused similar changes in all three salivary glands, but the parotid was the most extensively involved. In the submandibular and sublingual glands, the initial effects were confined mainly to the periphery of the glands. Studies of control animals indicated that the acetone diffused rapidly and caused an almost immediate "fixation" of the glandular tissue, which was blanched within a few seconds after the injection. The initial inflammatory reaction was rapidly followed by the removal of the dead tissue and regenerative activity in which fibroblasts played a dominant part. Initial edema was replaced by fibrosis. The less injured glandular areas dedifferentiated to an almost uniform ductlike system, which then underwent squamous metaplasia. Within a week after the injection of acetone, the cells of the dedifferentiated structures began to proliferate and form buds from which new acini were formed. The entire lesion which was produced as a reaction to the acetone was repaired in about 3 weeks, and the only evidence of previous damage was a slight fibrosis in the capsule of the gland. In the experimental animals, the DMBA had a toxic effect on the connective tissue, which inhibited the fibroblastic response produced by acetone alone. The DMBA crystals persisted for some weeks, and their toxic influence was evidenced by progressive acinar degeneration and dedifferentiation of acini, tubules, and ducts. There were also progressive vascular changes leading to swelling and hyalinization of their walls and narrowing of their lumina. No fibroblastic regeneration and little or no phagocytic activity occurred. The fibroblasts seemed to be more sensitive to the DMBA than the epithelial cells, and instead of removing the dead acinar tissue, as in the acetone controls, they died. Epithelial cells originating from collapsed, dedifferentiated and metaplastic glandular structures attempted to form an epithelial capsule around the "fixed" tissue, which was infiltrated by varying numbers of acute and chronic inflammatory cells. With further degeneration among the encysting epithelial cells the dead tissue remained undigested even 4 weeks after the injection of DMBA in acetone.

By about the 8th week, most of the dead glandular tissue was surrounded by an epithelial capsule, which appeared to link up with large distal parts of the ducts. Thus, sinuses or enlarged cysts containing exfoliated squamous cells, as well as dead glandular tissue, appeared. At this 8-week period, when the first tumors were seen, the vascular changes in the form of hyalinization of their walls and narrowing of their lumina was very conspicuous. Sarcomas, which appeared to arise independently of the cysts, were seen originating in the connective tissue surrounding the glands and also in the regenerating regions of striated muscle close to the glands. The DMBA in oil did not cause fixation of the muscle, so the toxic effects of the DMBA caused degeneration of muscle fibers, which was followed by the formation of rhabdomyosarcomas. Many rats had both carcinomas and sarcomas. Rats were considered at risk if they survived for at least 56 days, the approximate time when the first cancers were found; 41 males and 30 females were at risk in the experiments. The sex difference in induction time and cumulative rate of tumor formation for sarcomas was negligible.

When rates of tumor induction in male and female rats were plotted on various types of graphs, it was found that a straight line represented the cumulative incidence of carcinomas in males and females, but the graph for sarcoma incidence in both sexes was biphasic. While the initial slope for sarcomas paralleled that for carcinomas, the carcinoma slope was much shallower. The change to a biphasic type of graph occurred at about 140 days. In female rats given DMBA and acetone unilaterally, carcinomas appeared later and at a slower rate than in the males similarly treated. In the male animals, carcinomas appeared after a shorter latent period when the DMBA was given in acetone than when it was given in olive oil.

In essence, then, Glucksman and Cherry found that acetone caused fixation of the tissues which were rapidly removed by fibroblasts and macrophages. Extensive fibroplasia and squamous metaplasia of the remaining duct system preceded the complete regeneration of the gland, usually within 3 weeks. DMBA in acetone, and to a lesser degree in olive oil, caused necrosis of the glandular and surrounding tissue. Since the DMBA was toxic to fibroblasts and macrophages, it prevented the early resorption of the necrotic material, which remained undigested. It was encysted by epithelial outgrowth from neighboring glandular ducts, which had undergone squamous metaplasia.

The DMBA in acetone induced carcinomas and sarcomas earlier than the DMBA in olive oil, but the tumor incidence was the same for both groups. Carcinogenesis occurred in the regenerating epithelium of cysts and in regenerating connective and muscle tissue surrounding and encapsulating the glands. Rats surviving for 400 days still produced sarcomas, indicating that while the time rats were at risk for carcinoma development was limited, that for sarcoma persisted throughout the life of the animal. Sarcomas seemed able to appropriate the blood supply of neighboring carcinomas, to invade and strangulate them, and thus to cause their regression.

Because of the indication of a sex difference in rate of induction of carcinomas of the salivary glands in rats, Glucksman and Cherry[104] examined the process of carcinogenesis in intact and castrate male and female rats, as well as in animals treated with testosterone, with stilbestrol, or with testosterone plus stilbestrol. They also investigated the influence of L-thyroxine and of methylthiouracil on tumor induction in intact and castrate male and female rats. They were interested not only in whether sex and thyroid hormones influenced carcinogenesis, but, if so, at what stages and in which tissues and cells. A group of 521 black-hooded, laboratory-bred rats 2 to 3 months of age were used for the experiments. Of the group, 122 males and 115 females were surgically castrated 4 to 6 weeks before the injection of DMBA. Also, 112 intact and castrate male and female animals received no treatment other than the injection of DMBA in acetone into the right and left salivary gland complex. These animals were allowed to survive until clinical signs of tumors appeared. Thirty mg pellets of testosterone proprionate were inserted s.c. into 93 intact and castrate males and females. An additional 83 intact and castrate males and females had stilbestrol added to the drinking water at a daily dose of about 0.002 mg per rat; 79 intact and castrate males and females had testosterone pellets inserted s.c. and were given stilbestrol in their drinking water; 77 intact and castrate males and females had L-thyroxine (1 mg/1000 ml) added to the drinking water, giving a daily dose of approximately 0.02 mg for each rat. An additional 77 intact and castrate males and females received methylthiouracil (1 g/1000 ml) in the drinking water at a daily dose of approximately 0.2 g per rat.

The first carcinomas and sarcomas appeared after 40 days, the last carcinoma before 200 days, and the last sarcoma at about 500 days. The cumulative percentage of carcinomas rose linearly with time, but that of sarcomas was biphasic, as previously described. More carcinomas were elicited more rapidly in males than in females and castrated rats. More sarcomas were induced in male rats up to 200 days, but in later stages the sex difference and incidence was obscured by the tardy development of tumors in both sexes. Testosterone increased and accelerated the induction of carcinomas and sarcomas in females and castrates, whereas stilbestrol decreased and retarded tumor development in all but intact females. Thyroxine slowed the development of carcinomas and sarcomas in intact and castrate males and females and also lowered the percentage of carcinomas induced in intact and castrated females. Methylthiouracil decreased the incidence of carcinomas in intact and even more in castrate females, but had no significant effect on the rate of carcinogenesis in males and on the development of sarcomas in both groups of females. Testosterone and stilbestrol given simultaneously had an effect similar to that of testosterone alone.

Other studies by Glucksman and Cherry[103] concerned the effect of varying the carcinogenic dosage on the sex difference in tumor induction. They found that at low concentrations of DMBA (½ to 1%) twice as many sarcomas and carcinomas of the salivary glands were induced in male as in female rats. Additional estrogens reduced the number of neoplasms in males by one half, whereas testosterone doubled the number of neoplasms induced in females. The sex difference disappeared at higher dose levels of the carcinogen (2%). They also found that females were more sensitive than males to the toxic effect of DMBA, even though they were less sensitive to the carcinogenic action of the polycyclic hydrocarbons.

Quite apart from the differences in the incidence of cancers of accessory sex organs, such as the breast in men and women, epidemiological studies indicate that there are appreciable variations in the sex ratio for cancers of the mouth, pharynx, larynx, esophagus, stomach, lung, and bronchus, which are 2 to 20 times as frequent in men as in women. Epidemiological studies indicate that malignant disease of the salivary glands is a disease of older age groups and appears to occur about two times as frequently in men as in women. Some of these differences could be attributed to environmental factors such as occupation, social status, smoking and drinking habits, which might vary in incidence in different countries and within countries. There may be, however, some factors inherent in the male and the female which promote or inhibit the development of carcinomas at various sites. Glucksman and Cherry undertook their series of experiments because they believed that an understanding of these factors might help in the prevention or therapy of malignant disease. The volume and secretory activities of the tubules of the male rodent submandibular gland greatly exceed those of females. These differences diminish after castration or hormonal treatment.[105,106] The male submandibular gland of mice also produces more nerve growth factor,[107] although the secretion can be stimulated in the female by testosterone. Irradiation had been shown by Glucksman and Cherry[102] to injure the acini and induce regeneration from the secretory tubules of the submandibular and the intercalated ducts of the sublingual glands, which later resulted in the appearance of adenomas, and both these processes are more marked in males than in females. They suggested that the difference in action of the sex and thyroid hormones on tissues undergoing carcinogenesis was due to some general metabolic process rather than to a specific endocrine effect.

Several investigators[98,99,101] have suggested that when a carcinogen was administered in the form of a pellet, the resulting gradual absorption acted on the glandular components and produced squamous cell carcinomas, as demonstrated in guinea pigs, mice, and rats, but not in hamsters. Chaudhry and co-workers[108] undertook an investigation to determine if hamsters responded differently from other animals because of the dose level of the carcinogen, its mode of administration, or a species difference. Accordingly, the submandibular glands of 50 male albino rats and 50 male Golden Syrian hamsters were implanted with pellets containing 0.2, 0.4, and 0.8 mg of DMBA in polyethylene glycerol. All animals were sacrificed at the end of 24 weeks. The implantation of the different amounts of carcinogen in pellets into the submandibular glands of rats and hamsters resulted in different types of tumors in the two groups. More squamous cell carcinomas were produced in the rats, whereas more sarcomas were produced in the hamsters. This statistically significant difference might have been a result of species specificity, a hypothesis which the investigators point out needs further investigation. They also found that an increase in the concentration of DMBA resulted in a corresponding increase in the incidence of malignant tumors in both hamsters and rats. In the rats, 27% developed malignant tumors, four of which were sarcomas and eight, squamous cell carcinomas. In the hamsters, 25% developed malignant lesions, all of which were sarcomas. An increase in the carcinogen concentration was associated with a corresponding increase in the incidence of the sarcomas. No adenocarcinomas, collision tumors, or so-called mixed tumors of the salivary glands were noted in any of the animals.

Later Schmutz and Chaudhry[109] attempted to determine the most effective dose of the carcinogen that would produce a maximum tumor yield within a specified period of time without undue mortality. They were attempting to produce an experimental model for future work. Five groups of Sprague-Dawley rats 2 to 3 months of age were used. Each group of ten animals received a single injection — a measured dose of carcinogen — except those in Group V, which were injected with an additional 250 mg 8½ weeks after the initial treatment with 250 mg. Animals in Group I received 250 mg; Group II, 500 mg; Group III, 750 mg; Group IV, 1000 mg; Group V, 250 mg twice. The animals were killed at the end of 20 weeks. The incidence of induced tumors ranged from 38 to 45% in all groups except Group IV, in which all of the animals developed tumors. The induced tumors in all groups were squamous cell carcinomas except in three animals. One which received 750 mg developed a well-differentiated fibrosarcoma, and two which received 1000 mg showed the presence of both squamous cell carcinoma and fibrosarcoma. Rat strain specific responses to a carcinogen were studied in 90-day-old male Wistar, Sprague-Dawley, and Long-Evans rats by implanting pellets of 9,10-dimethyl-1,2-dibenzanthracene into the submandibular glands of 16 male rats of each strain.[110] Epidermoid carcinomas or stages which characteristically precede their formation were observed in all animals. Microscopic study revealed that while all tumors were essentially similar in the three strains, the Long-Evans rat developed carcinoma 3 weeks earlier than the Srague-Dawley and Wistar rats. All lesions which were permitted sufficient time to develop resulted in epidermoid carcinomas. The Long-Evans rat developed the first observed tumors at 7 weeks following the implantation of the carcinogen, whereas the Wistar and Sprague-Dawley rats required 10 weeks to develop tumors.

A comparison was made between the carcinogenic activities of benzpyrine and benzpyrine-7,8-dihydrodiol in adult C57 black mouse submandibular salivary glands.[111] A difference in the type of tumor induced by the two compounds was found. Benzo[a]pyrine (B[a]P) induced tumors of the salivary glands at the site of injection, whereas the dihydrodiol induced malignant lymphosarcomas, particularly of the thymus, which were often metastatic to other organs. It was postulated that the difference between the two compounds might be partly due to their relative water solubilities. B[a]P is lipid-soluble and would tend to stay at the site of injection, whereas the more water-soluble dihydrodiol may be dispersed and able to act on more sensitive target tissues. A consideration of the toxic effects of the compound and the resultant stimulation of regenerative cell division may provide an additional explanation for the two different sites of action of B[a]P and its dihydrodiol derivative.

Wigley and Carbonell[112] injected an emulsion containing two mg DMBA into the submandibular glands of young adult male C57 black mice. DMBA in acetone was emulsified in distilled water or in KY water soluble jelly. The subsequent induced preneoplastic changes, including duct hyperplasia and squamous metaplasia, were studied by both light and electron microscopy. Squamous cell carcinomas and fibrosarcomas arose with approximately equal frequency after 12 to 24 weeks. About 75% of the surviving animals developed one or both types of tumor. The specific cell type from which the squamous carcinomas developed was studied in depth. Both ultrastructural characteristics and histochemical markers of differentiated function were used as criteria of cell identification. They found that the main target cell in carcinogen-induced tumorigenesis was derived from the granular tubules. These cells continued to secrete characteristic proteases during the early preneoplastic phase, but that function was lost in advanced squamous metaplasia and carcinoma.

In a similar study, Kim and co-workers[113] used the rat submandibular gland and DMBA injection as a model system to study the process of carcinogenesis. One submandibular gland of male Sprague-Dawley rats was injected with 1 mg DMBA in liquid petrolatum. The other gland received the same amount of liquid petrolatum without DMBA and served as a control. Control and experimental animals were sacrificed at 2, 4, 7, 9, 14, 18, and 20 weeks after the injections. One of the most striking features noted early, following the injection of DMBA, was the appearance of ductlike structures which seemed to form by metaplastic transformation of secretory cells, mostly acinar cells. These structures were noted at about 2 weeks following the injection, well before signs of malignant change began to appear at about 14 weeks following injection. The fine morphological alterations seen in the cells of these ductlike structures and in acinar cells undergoing metaplastic transformation were characteristic of depressed protein synthesis. The secretory cells of the control gland did not reveal the same alterations, suggesting that the changes in the secretory cells of the experimental gland were related either directly or indirectly to the presence of DMBA and might represent an initial event in carcinogenesis. The origin of the metaplastic changes, however, remained unclear. There seems to be uncertainty regarding the origin of the metaplastic epithelium in salivary glands because detailed structural changes that occur in cells of carcinogen-injected glands have not been clearly described. It has been suggested that metaplastic changes arise from striated ducts, from intercalated ducts, or directly or indirectly from acinar epithelium. Kim et al. were not able to resolve the problem. As they pointed out, it was uncertain whether any of the changes seen in the injected glands were in fact related to the process of malignant transformation which took place at a much later stage. Their study was done well before any signs of malignant change developed. They reported that the structural differences between the control and experimental glands included the disappearance of parallel arrays of endoplasmic reticulum, decreased numbers of rough endoplastic reticulum, and increased numbers of free ribosomes in the cells of injected glands. These fine structural differences were consistent with the changes seen in other secretory cells associated with depressed protein synthesis. They also found a reduction in the size and number of the Golgi complexes and an empty appearance of the Golgi cisternae in the cells of the ductlike structures in experimental glands. These changes also reflected a depressed state of protein synthesis and secretory granule formation. The changes in structure and in DNA activity were related to the presence of DMBA in the salivary gland. Ductlike structures and the cells constituting these structures in the injected gland were similar to developing exocrine glands prior to the appearance of secretory granules. The changes observed following the injection of the rat submandibular gland with DMBA might represent a process of "dedifferentiation".

As pointed out by Wigley and Carbonell,[112] the regeneration process was thought to be essential to salivary gland carcinogenesis, since the mitotic rate in the gland is normally very low. They also observed that mitosis could be stimulated in rodent salivary glands by isoprenaline. Salivary gland tumors in rats could be induced by methylnitrosourea only when the animals were pretreated with isoprenaline.[90] The question then was whether isoprenaline would be effective in increasing the susceptibility of the mouse submandibular gland to polycyclic hydrocarbon carcinogenesis. Isoproterinol was injected i.p. three times a week for 2 weeks following a single injection of DMBA in olive oil into the salivary glands of male strain DD mice.[110] The salivary glands treated with carcinogens showed a relatively constant sequence of changes and developed epithelial and mesenchymal tumors irrespective of isoproterinol administration. In fact, isoproterinol not only failed to influence the development of another type of tumor, but it retarded the oncogenic process. The suggestion was made that the small increase in cAMP induced by isoproterinol may have retarded the development of experimental carcinomas in the mouse salivary glands.

Obviously, continued studies to determine the factors necessary for the chemical induction of malignant epithelial tumors in the salivary glands of experimental animals were needed. Chaudhry and Schmutz[115] studied the induction of submandibular gland tumors in hamsters under the influence of prednisolone and thalidomide. They found no significant difference in the incidence of induced fibrosarcomas in hamster submandibular glands when they received a single intraglandular injection of 100, 200, 500, or 1000 mg of prednisolone concurrently with the administration of the DMBA in liquid petrolatum; however, when 500 mg of prednisolone was injected i.p. every week throughout the experimental period, there was an inhibitory trend. Thalidomide failed to show any influence on induction of tumors when given to hamsters either as a single 5 or 10 mg dose or as a 5 mg injection every week.

When the male rat was shown to be an effective experimental animal for the study of salivary gland carcinogenesis, Shklar and his group, as well as others, used the model to study the effects of systemically administered pharmacologically active chemicals on salivary gland carcinogenesis. Anbari et al.[116] implanted DMBA into the submaxillary gland of male rats. Some of the animals were given daily 0.5 mg parenteral injections of cortisone acetate 2 weeks after the start of the experiments. Half of the animals were sacrificed 8 weeks after the start of the experiment. The other half were sacrificed 4 weeks later. The animals that received cortisone developed carcinomas more rapidly and in larger numbers than the DMBA controls. Although the difference could not be discerned at 8 weeks, by 12 weeks the DMBA cortisone group had developed frank malignancies in over 50% of the animals, whereas only 10% of the animals developed malignant tumors in the DMBA controls. Turbiner et al.[117] studied the effect of cold stress on chemical carcinogenesis by implanting DMBA into the submaxillary glands of male rats and placing them in a cold room maintained at $3°C \pm 1°$ for the duration of the experiment. Beginning at 8 weeks, two cold-stressed and two control room-temperature animals were killed at weekly intervals up to and including 19 weeks. Initially, the tumors in both the control and cold-stressed animals were comparable in size, but by the 13th week tumor growth in the cold-stressed rats exceeded that observed in the animals maintained at room temperature. At the time of death, the cold-stressed animals weighed 81 g less, on an average, than the animals in the control group, although they consumed larger quantities of food and water. In all cases, the tumors which developed in the implanted salivary gland were epidermoid carcinomas. The histologic appearance of the tumors was essentially similar in both groups. The number of mitotic figures, hyperchromaticity, and cellular dimensions were comparable. It was not possible to differentiate between a cold-stressed and a control animal tumor by means of its histologic appearance. However, an impression was gained that the sequence of events leading to neoplastic transformation had proceeded somewhat more rapidly in the cold-stressed animals than in their room-temperature-maintained controls. Following the neoplastic transformation, the tumors developed more rapidly and developed to larger size in the cold-stressed rats, but the cellular patterns were essentially similar in the experimental and control groups. The fact that extended cold stress produced an increased tumor size is not surprising in view of the fact that stress produces an increase in respiratory exchange and food intake as well as metabolism. It was postulated that the increased metabolism resulting from the prolonged cold stress was responsible for increasing the growth rate of the tumors. It was also suggested that cold stress might act to stimulate tumor growth by depressing antibody formation. The fact that Shklar's group had already shown that cortisone increased the susceptibility of rats to DMBA was not mentioned, although one would suspect that cold-stressed animals produced more adrenal cortical activity than nonstressed animals.

Sheehan and Shklar[118] investigated the effect of cyclophosphamide on salivary gland tumors. Pellets of DMBA were implanted into the right submaxillary glands of male and female albino rats. One group of these animals received cyclophosphamide (cytoxan) biweekly i.p. In the DMBA-cyclophosphamide animals, epidermoid carcinomas developed more rapidly and were histologically more anaplastic than those in the DMBA controls. In addition to the epidermoid carcinomas, fibrosarcomas and adenocarcinomas developed in the DMBA cyclophosphamide animals. This augmentation of salivary gland carcinogenesis was interpreted in terms of the depression of the immune response by the cyclophosphamide. In another series of experiments, Shklar and Sonis[119] studied the effect of methotrexate on experimental salivary gland neoplasia in rats. They implanted pellets of DMBA in the submaxillary glands of male Sprague-Dawley rats 90 days of age. Some of the animals then received s.c. injections of 1 mg sodium methotrexate twice weekly. The methotrexate was found to accelerate the process of chemical carcinogenesis in the rats' submandibular gland. The carcinomas developed within the walls of epidermoid cysts as in the DMBA animals, but were well established at 10 weeks, whereas the carcinomas were just beginning to develop in the control DMBA animals. At 12 weeks, the carcinomas in the DMBA plus methotrexate animals were considerably larger than those in the DMBA group, and there was a greater invasiveness into the glandular tissue. Shklar and Sonis postulated that the systemically administered methotrexate acted to depress the animals' immune response to the developing carcinomas, thus permitting the more rapid growth and deeper invasion. However, unlike the effect of the cyclophosphamide-DMBA treatment of animals, methotrexate resulted in epidermoid carcinomas in all cases, while the cyclophosphamide not only augmented carcinogenesis, but resulted in the formation of sarcomas and adenocarcinomas, as well as epidermoid carcinomas. Another study provided further evidence that an immunosuppressive agent enhanced the action of carcinogenic chemicals. Since lowering of immune response appeared to enhance carcinogenesis, it was only logical that Shklar and his group should study the effects of a nonspecific enhancer of cell-mediated immunity on the development of chemically induced salivary gland tumors.

Pisanty et al.[120] therefore investigated the effect of levamisole. They injected 0.6 mg levamisole once daily on three consecutive days at intervals of 2 weeks into animals that had DMBA implanted in their submandibular glands. DMBA animals that were not injected with levamisole developed tumors more rapidly, were more anaplastic, and showed greater invasiveness than those which developed more slowly in levamisole-injected animals. The researchers concluded that levamisole, while it did not prevent carcinogenesis from taking place, delayed the process of carcinogenesis by approximately 3 weeks. As tumors observed in levamisole-injected animals were less invasive and relatively more papillary than those in the nonlevamisole injected animals, they postulated that the levamisole reduced the severity of the carcinomas and retarded neoplastic transformation. They believed that this reaction was a result of the nonspecific cell-mediated immune enhancement induced by levamisole on the immunologic system of the animals studied.

Other studies on general metabolic influences on salivary gland carcinogenesis have been reported. Ciapparelli et al.[121] studied the effects of zinc on salivary gland tumors by adding 50, 100, and 250 ppm zinc to the deionized drinking water (analar zinc sulfate). Two weeks after the animals were placed on either the deionized drinking water or the water containing various amounts of zinc, pellets of DMBA were implanted into their submaxillary glands. The animals were sacrificed 3 months after implantation of DMBA. The tumor growth induced in the submandibular glands of the albino rats was retarded when the zinc concentration in the drinking water reached 250 ppm. In the group ingesting the deionized drinking water, well-differentiated squamous carcinomas developed. As the zinc concentration of the drinking water increased, the carcinomatous epithelium became progressively less and the inflammatory response more marked. The authors were unable to explain the inhibitory effect and the change in pathogenesis produced by the zinc in the drinking water, but it is obvious that additional research in this area might lead to ways to control malignant disease.

In another study on the effects of nutritional, endocrine, and thermal variables as possible carcinogenic cofactors, Rowe et al.[122] studied five groups of Sprague-Dawley rats. The groups were divided into vitamin A deficiency, hypercortisonism, hyperthyroidism, hypercortisonism plus hyperthyroidism, and a "windy, cold, humid environment". The submaxillary salivary glands of each animal were implanted with DMBA pellets as the animals were placed on the experimental protocol. Animals were sacrificed at 14 weeks following the implantation. Of the five regimens, only vitamin A deficiency increased the number of malignant epithelial neoplasms. No significant yield difference was found among the other four groups. It is regrettable that these investigators did not discuss their findings in the light of those reported by Shklar's group, but there is obviously a discrepancy which needs further investigation.

Englander and Cataldo[123] compared tumor induction time and the type of tumor formed in rat submandibular glands implanted with DMBA which had been ligated 5 days prior to the implantation. Results showed no difference in tumor induction time or the type of tumor formed. The ligated controls showed degeneration and atrophy of the affected gland and replacement of glandular tissue by scar tissue. The DMBA controls followed the expected pattern, with all rats eventually developing well-differentiated squamous cell carcinoma. The DMBA-ligated rats showed a combination of the two processes, that is, scar tissue formation and carcinogenesis, each acting separately from the other. There was no difference in induction time or tumor anaplasticity between the DMBA control rats and the DMBA-ligated rats. The investigators postulated that only factors relating to a change in the immunologic response of the host affect the latent period of tumor induction in submandibular gland carcinogenesis.

Almost all of the experimental inductions of salivary gland tumors by polycyclic hydrocarbons have been done by the implantation of pellets of the carcinogen or by injection of the carcinogen directly into the submaxillary gland. It seemed of interest to Sela and co-workers[124] to investigate the effect of placing the carcinogen into the salivary gland by way of its main excretory duct. Using this method, they expected that the carcinogen would act on all parts of the glandular epithelium, from that of the large ducts to that of the lobules. Previous work by this group had demonstrated that cannulation of the parotid ducts of rats followed by the introduction of material into the glands resulted in a distribution of material throughout the glandular parenchyma and eruption of the material into the connective tissue stroma of the gland. In these experiments, the investigators injected DMBA in water and olive oil into Sabra strain rats via the duct of the parotid gland. The animals were killed at intervals up to 60 weeks. Squamous metaplasia of the large salivary gland ducts appeared in a few animals before the 16th week. Animals killed 16 weeks after the last dose showed fibrosarcoma and one epidermoid carcinoma. In general, most of the tumors produced were fibrosarcomas with one squamous carcinoma. No pleomorphic adenomas occurred. The majority of the tumors appeared approximately 100 days following the intraductal injection of the chemical carcinogen. The salivary gland carcinogenesis model appears to be an exciting development for the study of carcinogenesis in general and for the study of the pathogenesis of different kinds of salivary gland tumors specifically.

HAMSTER CHEEK POUCH

Normal Cheek Pouch

The hamster cheek pouch is a very interesting and special organ. Many investigators have referred to it as "an oral mucous membrane". While it is not an oral mucosa per se, the hamster cheek pouch has provided a very important model for the study of carcinogenesis and malignant growth. The cheek pouch is lined by a membrane two or three epithelial cells thick, which covers loose fibrous connective tissue. Unlike skin and oral mucous membranes, the pouch has no adnexae such as pilosebaceous, mucous and/or salivary glands. In addition, the mucosa of the cheek pouch of the hamster has been found to be an immunologically privileged site in that homologous or heterologous transplants are readily accepted in an untreated animal.[125] This immunologically privileged status is said to result from the paucity of lymphatic drainage and the presence of an areolar connective tissue barrier beneath the pouch mucosa. The immunological privilege is not permanent. Shepro et al.[126] demonstrated that the superior cervical lymph node will produce an immune response, but only after a sufficient time has elapsed for the diffusion of antigen to take place. The delay in that response allowed a transplant to survive long enough to become established and studied. Recently, a finding of great importance to the study of experimental carcinogenesis was described by Ferguson and Smillie.[127] They demonstrated that the wall of the normal hamster cheek pouch possessed small nodules, the histologic structure of which closely resembled that of hyperplastic epithelium. Their finding was significant in view of the number of reports that described nodules on the cheek pouch wall as one of the early results of carcinogen application. For example, Woods and Smith[128] stated that "nodules and overt tumors were sampled at random." A failure to appreciate the presence of nodules as a normal feature of the pouch wall of the hamster could lead to the interpretation that they resulted from carcinogen application, since the histologic details of the normal nodules are comparable to those of carcinogen-induced epithelial hyperplasia. It seems that the normal physiology and structure of the hamster cheek pouch is in need of a study in depth and might provide a fruitful field for further investigations.

Experimental Carcinogenesis

The literature through 1971 has been reviewed by Homburger (242 references),[129] Homburger,[130] and Shklar.[131,132] They both pointed out that the experimental pathology of "oral" mucosal carcinoma actually began with the investigations of Salley,[133] who first demonstrated that epidermoid carcinomas could be produced in the buccal pouches of hamsters by multiple paintings with chemical carcinogenic agents such as dimethylbenzanthracene (DMBA). Salley's work was important because it helped to establish the hamster cheek pouch as a viable model for the study of carcinogenesis. He pointed out[134] that there were four distinct histologic phases during the process of carcinogenesis: inflammation, degeneration, regeneration, and hyperplasia. The relationship of age, sex, and concentration of chemical carcinogen to the production of malignant tumors in the hamster cheek pouch was studied later by Morris.[135] He demonstrated that the optimal concentration of DMBA for high tumor yield and short latent period was 0.5%. Lower concentrations produced tumors more slowly, and higher concentrations, such as 1.5%, were toxic. He found that the application of a carcinogen three times a week was more effective than twice a week, but that a smaller total dose was required to produce tumors in animals when the carcinogen was given only twice a week. There were apparently no sex differences in the response to DMBA, but age was important. Hamsters aged 18 months were more resistant to tumor formation by the application of DMBA than were younger ones, up to 9 weeks of age. The details of the story of cheek pouch carcinogenesis with DMBA is well covered in the reviews referred to above.

The hamster cheek pouch model has been used to demonstrate that DMBA carcinogenesis could be augmented by chronic irritation of the pouch mucosa[136,137] as well as by systemic factors. It was found, for instance, that vitamin A deficiency augmented chemical carcinogenesis following the application of DMBA to the hamster cheek pouch. However, Homburger[138] showed that when 10% vitamin A palmitate was applied to the pouch after the carcinogen treatment, infiltrating squamous cell carcinoma resulted, whereas when 0.5% liquid paraffin was applied after the carcinogen, small papillary tumors 1 to 2 mm in diameter, which showed foci resembling intraepithelial carcinoma, regressed or reverted to benign papillomas within 2 months. Cavalaris et al.[139] also demonstrated that vitamin A had an enhancing effect on DMBA-induced tumorigenesis. They demonstrated that both premalignant and malignant changes were more numerous and more extensive in a vitamin A treated group of hamsters than in a group treated with only DMBA for the same period of time. While the role of vitamin A in enhancing DMBA-induced tumorigenesis was not explained, the effect might have been related to proliferation caused by vitamin A on the cheek pouch epithelium and/ or its labilizing effect on cellular membrane systems. The effect of vitamin A on epithelial differentiation and keratinization has been well established. Dresser[140] showed that vitamin A might act as an adjuvant and might also stimulate lymphocyte proliferation. Levij and Polliack,[141] who had also noted the appearance of dense mononuclear cell infiltrates in the lamina propria of the hamster cheek pouch mucosa following the application of vitamin A, further studied this phenomenon. They treated the hamster cheek pouch with 20% vitamin A palmitate in liquid paraffin for 12 weeks and noted dense mononuclear cell infiltrates in the lamina propria and muscular wall. These infiltrates often appeared nodular and consisted of lymphocytes and large irregular mononuclear cells similar to those found in malignant lymphomas. Although the dense mononuclear infiltrates were reminiscent of the lymphoreticular hyperplasia seen in malignant lymphomas, the lesion was diagnosed as a "simulated lymphoma" or a "lymphomalike" lesion. More recently Shklar et al.[142] reported that carcinogenesis in the hamster cheek pouch was inhibited by the feeding of nontoxic doses of the synthetic retintoid 13-*cis*-retinoic acid. They painted the buccal pouch of the hamster with DMBA in heavy mineral oil and fed the animal some 10 mg of 13-*cis*-retinoic acid in peanut oil twice a week. The carcinogen and retinoid were administered on alternate days. Those animals which received the 13-*cis*-retinoic acid exhibited a significant delay in DMBA carcinogenesis.

Assuming that there was a competition between DMBA and vitamin D for active sites on DNA, Rubin and Levij[143] prepared solutions of DMBA in combination with vitamins D_2 and D_3. Local application of the two vitamin Ds with DMBA caused inhibition of DMBA tumor formation in the hamster cheek pouch. The inhibition by the simultaneous topical administration of vitamin D was thought to be nonspecific. They postulated that the control of the mitotic process was decreased because there was a decreased amount of calcium bound to the membrane of the tumor cells when vitamin D was applied, leading to increased cellular permeability with lack of controlled flow of metabolites to and from the cytoplasm. Vitamin D was believed to cause the induction of the synthesis of an alkaline phosphatase/calcium binding protein, which made the calcium unavailable to the tumor cell membrane. It is obvious that a considerable amount of work in this area still needs to be done.

Homburger[130] warned that many changes induced by DMBA in the hamster cheek pouches are actually inflammatory reactions which may be morphologically difficult to separate from malignancy, but are actually biologically benign. He suggested that it was wise to view with caution the conclusions drawn from relatively short-term DMBA applications to cheek pouches, because during the early phases, which last 2 to 4 months, the neoplastic and inflammatory processes are intertwined. He pointed out that at about 5 months following DMBA application, clearly neoplastic lesions emerged and inflammatory changes subsided. Eisenberg[144] undertook a study to determine whether continuous application of DMBA was in fact necessary for the production of frank carcinoma once histologic changes consistent with epithelial dysplasia were seen. She painted the buccal mucosa of hamsters three times a week with DMBA in mineral oil and sacrificed animals while on the treatment regimen after 8, 10, 11, 13, and 15 weeks. Gross changes of the pouch were evident by 10 weeks. In another group of animals, the carcinogen application was discontinued after the 10-week period. In this second group of animals, extensive involvement of the cheek pouches was seen both grossly and histologically by weeks 13 and 15. Once the process of malignant transformation by DMBA was established, the continuous application of a carcinogen became elective, since 100% of the experimental animals which showed epithelial dysplasia at 10 weeks showed frank carcinoma by 13 and 15 weeks, even though the application of carcinogen had been discontinued after 10 weeks.

In studying a variety of cocarcinogenic influences on neoplastic induction by DMBA in the cheek pouch of the hamster, Shklar[145] found that systemically administered cortisone resulted in more extensive spread and invasion of chemically induced buccal pouch carcinoma. Since the corticosteroid might have acted to depress an immunologic response to the developing malignancy, Giunta and Shklar[146] followed up the experiments using specific hamster antilymphocyte serum (ALS). Animals injected with ALS showed a more rapid development of widespread buccal pouch lesions induced by applications of DMBA. The lesions in the ALS animals were also more anaplastic and more deeply invasive than in controls. When cortisone was topically administered during DMBA carcinogenesis in the hamster cheek pouch, inhibition of tumor formation resulted.[147] Whether this was due to the inhibition of mitotic activity by cortisone or something more specifically related to the DMBA cheek pouch model was not clear. Gillman et al.[148] found no suppression by cortisone of mouse skin tumors induced with 3-methylcholanthrene. On the other hand, other experiments showed that cortisone did inhibit mouse skin tumors when the skin was painted with DMBA,[149] methylcholanthrene, or benzopyrene.[150,151] The results reported by Polliack and Levij,[147] who gave cortisone and DMBA simultaneously, were quite different from those reported by Sabes et al.[152] They demonstrated that DMBA-induced hamster cheek pouch carcinogenesis was enhanced when cortisone was applied topically 3 min prior to the application of the carcinogen. However, the reverse procedure, that is applying the cortisone 3 min after the DMBA application, resulted in no significant difference in the incidence of tumors as compared with the controls. The divergence of the results may have been related to the difference in the dose, since Polliack et al. had used a 0.05% cortisone acetate solution, and Sabes had used a 0.025% solution.

The contradictory reports concerning the effect of cortisone on experimental carcinogenesis might have been related to differences in the route of administration, dosage, time of administration of the steroid in relation to the application of the carcinogen, duration of treatment, animal species, and target organs. In any event, this is another area that requires further exploration to which the oral biologist could make an important contribution.

Because hormonal factors were known to play an important role in the induction and development of some experimental tumors, Levij et al.[153] and Polliack et al.[154,155] utilized the hamster cheek pouch to study the effects of estrogen and testosterone on chemical carcinogenesis. They castrated a group of male hamsters. In some, estrogen was injected while the cheek pouch was treated with DMBA; in some, testosterone was injected, and in others nothing was injected during the treatment. Testosterone was also injected into intact noncastrated male animals. The cheek pouches of all animals were painted with DMBA for up to 12 weeks. Animals were examined at 9 and 12 weeks. Levij and his group found that estrogen, when administered i.m. to castrated male hamsters during the topical application of DMBA to cheek pouches, enhanced the development of malignancy. The promoting effect of the hormone was marked after 12 weeks, but was not present after 9 weeks. In noncastrated animals, estrogen did not promote carcinoma formation at any stage of the experiment. Intramuscular injection of testosterone proprionate into intact animals receiving DMBA resulted in fewer carcinomas. The tumor incidence in intact and castrated animals treated with DMBA alone was similar throughout the experiment. The relationship of hormones to chemical carcinogenesis seems to be another area for further research by the oral biologist.

Shklar et al. and Sheehan et al.[157] pointed out that hamster cheek pouch carcinogenesis was augmented by the systemic administration of such antimetabolites as methotrexate or azathioprine. The effect of systemically administered antimetabolite drugs in augmenting chemical carcinogenesis was of great interest and opened many avenues for investigating metabolic relationships during chemical carcinogenesis. When administered systemically, 5-fluorouracil (5 FU)[158] exerted an effect similar to that of methotrexate resulting in neoplasia which involved the entire buccal pouch. The lesions rapidly invaded the underlying tissues, and the neoplastic pouch became fixed and could not be everted. On the other hand, Weathers and Halstead[159] observed that the daily application of a 1% solution of 5 FU in propylene glycol did not incite a histologic change in the normal hamster pouch mucosa and provided some evidence that 5 FU delayed development of carcinomas from DMBA-induced dysplasia. Since this was a pilot study, they recognized the need for further studies in which different concentrations and more frequent applications of 5 FU would be applied to the cheek pouch during chemical carcinogenesis.

The need for an experimental model in which to study tobacco and betel nut carcinogenesis had long been sought. It was thought that the hamster cheek pouch might provide a good model for that purpose. Suri et al.[160] extracted betel nut and tobacco with dimethylsulfoxide (DMSO) and painted the extract on the mucosa of the hamster buccal pouch three times a week for 21 weeks; 90% of the hamsters developed leukoplakia. The extracts of betel nut resulted in the development of tumors in 38% of the animals. Application of the extract prepared from a mixture of betel nut and tobacco resulted in the development of tumors in 76% of the group. Extracts from tobacco alone caused leukoplakia in 66%, but no tumors were observed in the animals. The data indicated an early appearance of tumors in the hamsters receiving an extract made from the combined betel nut and tobacco mixture, and all tumors developed by the end of 18 weeks of painting; however, the incidence of leukoplakia was still increasing at the end of the experimental period. The authors pointed out that DMSO solubilized so many compounds that it was not possible to make a reasonable speculation about the identity of the active carcinogenic material in betel nut. It seems likely that it was fairly potent, since it was effective in producing lesions in spite of the simplicity of the method used to prepare the extract. Other studies have shown that tobacco smoke and tobacco itself did not produce malignant changes in the cheek pouch of the Syrian hamster.[161-165]

Because there is some clinical evidence that a relationship exists between the incidence of oral cancer and the consumption of alcohol, Elzay undertook a study to investigate the effect of alcohol and cigarette smoke as promoting agents in hamster pouch carcinogenesis.[166] The pouches of hamsters were painted with DMBA three times a week for 4 weeks. After the tumor initiating period, one group of hamsters was painted with 50% ethyl alcohol three times a week and received, by means of a smoking machine, whole smoke from half a king-size cigarette daily. Another group was painted with only alcohol three times a week, and a third group received only whole smoke daily. A fourth group of animals received only the DMBA application. A fifth group (no DMBA pretreatment) had their pouches painted with alcohol three times a week and received smoke daily, and a sixth group (no DMBA pretreatment) received only smoke daily. The hamsters were killed 175 days after the experiments began, and the pouches were fixed in formalin and prepared for microscopic examination. Malignant lesions were confined to the pouches of those animals which received DMBA only or DMBA plus alcohol and smoke. Parabasilar budding was noted in 88% and dyskeratosis in 81% of the pouches of those animals that received DMBA, alcohol, and smoke. The incidence of carcinoma *in situ* was 38% in that group. Animals which received DMBA and alcohol exhibited parabasilar budding in 60% of their pouches and dyskeratosis in 80%; 40% of the pouches exhibited either carcinoma *in situ* or carcinoma. This finding was similar to that noted in the animals which received DMBA, alcohol, and smoke; 80% of the pouches in the DMBA and smoke group showed parabasilar budding, whereas dyskeratosis was evident in all of the pouches of this group; 70% of the pouches displayed carcinoma *in situ* or carcinoma or both. Only 46% of the pouches in the DMBA-only group exhibited parabasilar hyperplasia or dyskeratosis, and 31% exhibited carcinoma *in situ*. Those pouches which received alcohol only were devoid of any changes. Hyperkeratosis was noted in the group which received smoke only. Thus, alcohol or smoke alone did not produce histologic change suggestive of premalignant or malignant epithelial transformation. Under the conditions of this study, alcohol and smoke acted as promoters, not as cocarcinogens or carcinogens. Elzay's results lend credence to previous clinical observations that heavy drinking and smoking appear to be related to an increased incidence of oral cancer.

Later Friedman and Shklar[167] studied the effect of alcohol ingestion on hamster buccal pouch carcinogenesis. In addition to painting the cheek pouch with DMBA in oil, they provided ethyl alcohol as a 10% solution instead of drinking water. They were thus able to study the effect of alcohol ingestion on the development of leukoplakia and oral carcinoma in a well-developed model system. Leukoplakia and epidermoid carcinoma developed more rapidly in those animals painted with DMBA and ingesting alcohol than in those painted with DMBA and ingesting water. The tumors were larger, more extensive, and more anaplastic. Since alcohol itself is not carcinogenic in the hamster system, its action was considered to be that of a cocarcinogen. The DMBA animals drinking alcohol developed dysplastic lesions and carcinomas some 2 weeks ahead of the water-drinking DMBA controls. Thus, the hamster cheek pouch model for the study of the effects of alcohol and other substances on carcinogenesis is proving to be an important bioassay system.

Using the hamster cheek pouch as a practical bioassay system for testing the carcinogenic effects of betel quid chewing, Ranadive et al.[168] made an effort to design experiments closely simulating the situation in the human mouth of either casual chewers or habitual chewers. They tested each betel quid ingredient (betel nut, betel leaf, lime, catechu, and tobacco) separately and in various combinations. To simulate the casual chewer, they painted the cheek pouch three times a week with aqueous extracts of the various test materials. To simulate the habitual chewer, they devised a more elaborate program. Test material was placed in wax pellets or in gelatin capsules. The pellets, the capsules, or the natural materials were introduced into the clean cheek pouches of the hamsters, which were then collared by ordinary insulating wire tied around their necks in order to keep the pellets and other materials in place. Fresh pellets were replaced every 2 weeks. Various betel quid ingredient combinations induced oral lesions ranging from massive atypia and precancerous lesions to frank carcinomas. Maximum lesions were observed in the groups receiving betel nut, lime, and tobacco combinations as well as in the polyphenol fraction of betel nut, which contains tannins. The mode of administration of the test ingredients determined their action and the extent of their effect on the cheek mucosa. Triweekly painting of extract of betel nut induced lesions in only 14% of the animals, whereas the same betel nut powder in gelatin capsules placed in the cheek pouch resulted in lesions in nearly 58% of the animals. In those animals with continuous contact with pieces of betel nut in the pouch, 63% produced lesions. The painting of the cheek mucosa with a polyphenolic fraction of the betel nut increased the incidence of oral lesions over that of the whole nut crude extract. When animals were painted three times a week with the betel nut and tobacco extract combination, the incidence of oral lesions was less than 22% whereas in the replaceable capsule group, addition of lime to the nut-tobacco mixture increased the incidence to over 58%. Direct exposure of the cheek pouch to betel nut pieces and tobacco resulted in lesions in about 53% of the animals. The fact that the whole betel quid extract induced fewer lesions than did the betel nut and the betel nut-lime extract suggested to Ranadive and co-workers that the betel leaves afforded some protection against oral carcinogenesis. Whether or not catechu in the quid provokes or inhibits carcinogenesis is not yet known; however, these workers were able to confirm the specific carcinogenicity of certain quid ingredients like betel nut and its combination with tobacco and lime and to show a direct relationship between frequency of chewing and induction of cancer. They also noted that a large number of animals (50 to 84%) developed lesions in the esophagus and the stomach. Their experiments suggested that chewing habits, besides causing oral lesions, might also be related to the increased esophageal and stomach lesions found in Kashmiri and other Indian populations.

When Nas, composed of tobacco 45%, lime 8%, ash 30%, plant oil 12%, and water 5%, was introduced into the buccal pouch of six consecutive generations of hamsters, tumors did not occur at the site of administration in any of the animals. There were, however, some histologic changes. In the early stages of the experiment (4 to 6 weeks after administration began), there seemed to be a disorientation of the cells of the basal layer of epithelium. Animals that had received the treatment for over 30 weeks showed degenerative and proliferative changes in the pouch epithelium, ranging from atrophy or an irregular diffuse hyperplasia of the epithelium to the formation of foci or proliferation consisting chiefly of basal cells. Changes in the submucous layer of the wall of the buccal pouch of the experimental hamsters were similar to changes observed fairly often in the mouths of persons taking Nas or betel for a long time, a condition which has been called submucous fibrosis.[169] Obviously the story of the relationship of tobacco to oral cancer is not yet complete, and further work is required.

The use of the hamster cheek pouch to bioassay for the presence or absence of carcinogens in various plant materials suspected of affecting development of esophageal cancer in man was adapted by Dunham et al.[170] The materials tested, often with calcium hydroxide added, were nine plants from Curacao used in native teas and remedies, a tobacco used in snuff, and an alkaloid (arecoline) which is present in betel quid. The plants were extracted in various ways. For instance, the plant material used as tea was boiled in distilled water for 30 min. Other materials such as arecoline were diluted with distilled water or with DMSO. Lesions developed in the cheek pouches of hamsters which were painted with the extracts of the plants *Annona muricata* (arecoline), *Acacia villosa* (a tea), *Heliotropium ternatum,* and *Krameria ixina*. The lesions were superficial spreading carcinomas. From the histologic appearance and the distribution of the lesions in the cheek pouch, Dunham and co-workers presumed that the test materials were weaker or slower in action than such other chemical carcinogens as DMBA. Their work further supported the view that the cheek pouch is valuable as a bioassay system.

As mentioned earlier, an important discussion of the premalignant lesions of carcinogen-treated hamster cheek pouches is to be found in the paper by Ferguson and Smillie.[127] They pointed out that the wall of the normal hamster cheek pouch possesses small nodules whose histologic structure closely resembles that of hyperplastic epithelium and that these nodules could be misinterpreted as a product of carcinogenesis. They were not able to find a reference in the literature to the small nodules observed by them as normal features of the cheek pouch wall, although Salley,[133,134] Morris,[135] and many others described nodules on the cheek pouch wall as one of the early results of carcinogen application. Some investigators sampled similar nodules and described their ultrastructural features as those of premalignant change. They quoted Woods and Smith[128] as stating that "nodules and overt tumors were sampled at random," and pointed out that failure to appreciate the presence of nodules as a normal feature of the pouch wall of the hamster could lead to their being interpreted as results of carcinogen application. The histologic details of the normal nodules were comparable to descriptions of carcinogen-induced epithelial hyperplasia reported in studies in which random sampling of cheek pouch tissue was done. Ferguson and Smillie showed that progression of premalignant lesions was associated with clinically visible vascularity. The progression of the static lesions that failed to culminate in cancer was not related to visible vascularity. The results of their studies suggest that the development of clinically obvious vascularity in a static lesion is a strong indicator of its likely progression to malignant disease. They postulated that the vascularization observed in their study could well be related to the elaboration of the tumor angiogenesis factor (TAF) described by Folkman et al.[171]

An investigation into the capacity of hamster cheek pouch static lesions to induce angiogenesis would be a logical next step in developing the hamster cheek pouch model because the existence of such a capacity would strengthen the possibility of developing an assay for impending malignant change in suspected premalignant lesions. Ferguson and Smillie also discussed the regression of lesions observed in their study. Either there was a temporarily decreased size in some lesions that were subsequently shown to be carcinomas, or there was a complete disappearance of lesions that had exhibited clinical features strongly suggestive of carcinoma. Their finding with respect to studies of chemical carcinogenesis in the cheek pouch was significant, since the tumor yield and latency period used as criteria in previous studies would be affected by the noninclusion of lesions that regressed.

Our conclusion is that despite the large amount of work that has already been done on the pathogenesis of carcinogen-induced hamster cheek pouch carcinoma, considerable room for investigative research remains.

Immunological Influences on Buccal Pouch Carcinogenesis

Although the hamster cheek pouch was considered to be an immunologically privileged site,[125] the evidence indicating that the privilege is not long lasting has already been discussed. Additional evidence that the cheek pouch is not immunologically privileged is accumulating. Mohammad and Mincer[172] induced dinitrochlorobenzene hypersensitivity in hamsters via the abdominal skin. When the allergen was then applied to the cheek pouch, local gross and histologic characteristics indicative of delayed hypersensitivity were observed. They thus illustrated that the buccal pouch of the Syrian hamster was capable of manifesting contact hypersensitivity to DNCB. Later Mohammad[173] studied hamster cheek pouches sensitized with the allergen DNCB either before the initiation of DMBA tumorigenesis or by direct application to already developed tumors. Sensitization prior to application of DMBA appeared to delay the development of tumors. However, only 52% of the hamsters treated with DNCB showed no evidence of tumor development or delayed tumor development. This was in contrast to the studies by Eisenberg and Shklar,[174] Cottone et al.,[175] and Shklar et al.,[176] who showed that levamisole, a nonspecific enhancer of cell mediated immunity significantly retarded chemical carcinogenesis of the hamster buccal pouch. Shklar and associates thought these observations offered further support to the concept that some defect in the host's immune response permitted the development of malignant neoplasms and that enhancement of the host's immune response retarded or prevented tumor formation. The inhibition of tumor progression reported by Shklar and co-workers was almost universal among the experimental animals. Thus, Shklar's findings seem at variance with Mohammad's. However, the proportion of DNCB-treated animals reported by Mohammad[173] to have an antitumor response is, in reality, in accord with the findings reported by Mohammad and Mincer[172] that topical DNCB application resulted in histologically demonstrable hypersensitivity in only 52% of Syrian hamsters. Moham-mad's study is at variance with those of other workers in another way. Even among the control animals receiving only DMBA, tumors developed at varying rates, some showing massive tumors and others only minimal changes after 12 weeks. Such varia-tions were possibly due to susceptibility or resistance of the individual hamster in-volved, or to a strain difference between the animals utilized by Mohammed and his group and those by Shklar and his group. The reason why the hamster cheek pouch is a good model for cancer research is because there is usually almost 100% tumor pro-duction following painting with DMBA. When there is an exception, reasons must be carefully sought. Marshack et al.[177] studied the effect of DNCB sensitization of the buccal pouch prior to or after DMBA tumor induction. When hamster cheek pouches were sensitized to DNCB following DMBA treatment, only benign noninvasive papil-lomas were formed. When sensitization with DNCB preceded DMBA tumor induction, squamous cell carcinomas *in situ* resulted. Treatment with DNCB prior to DMBA tumor induction was further effective in tumor prevention when DNCB rechallenge of the mucosa bearing tumors was employed since this resulted in most pouches becoming tumor-free. Although a few tumors were produced in the presence of a cell mediated response, a significant reduction in tumor invasiveness was shown to occur.

Elzay and Regelson[178] tested the effects of pyran copolymer on DMBA carcinoge-nesis in the hamster. Pyran copolymer is an immunopotentiator that induces inter-feron, stimulates lymphocyte response, and activates splenic and peritoneal macro-phages. The cheek pouches of one group of hamsters was painted with DMBA and received injections of pyran copolymer three times a week for 11 weeks. Another group of hamsters were painted with DMBA for 11 weeks, but did not receive the copolymer injections until the pouch lesions were clinically noted. All hamsters were observed for 4 weeks subsequent to the 11-week experimental period. Squamous cell carcinoma de-veloped in 83% of the hamster pouches that received DMBA alone. Carcinoma devel-oped in only 22% of the hamsters receiving pyran copolymer and DMBA simulta-neously. There was a slightly higher incidence of tumor yield in those hamsters receiving pyran copolymer after clinical evidence of pouch tumors developed. This finding was interpreted as indicating that pyran copolymer exerted its inhibitory action at the initiation phase of carcinogenesis. The mechanism of the inhibitory effect was not clear.

Again we have ample indication of the exciting opportunities open to the investigator of oral neoplasia. The hamster cheek pouch is an excellent model system for studying a wide variety of concepts and hypotheses related to oral carcinogenesis. Its special characteristics are awaiting exploitation by the investigative oral biologist.

REFERENCES

1. Pindborg, J. J., *Oral Cancer and Precancer,* John Wright and Sons Ltd., Bristol, 1980.
2. Moore, R. A., *A Textbook of Pathology,* W. B. Saunders, Philadelphia, 1944, 163.
3. Spitzer, W. O., Hill, G. B., Chambers, L. W., et al., The occupation of fishing as a risk factor in cancer of the lip, *N. Engl. J. Med.,* 293, 419, 1975.
4. Hirayama, T., An epidemiological study of oral and pharyngeal cancer in Central and Southeast Asia, *Bull. W.H.O.,* 34, 41, 1966.
5. World Health Organization, *World Health Statistics Annual, 1973-1976,* WHO, Geneva, 1976.
6. Russell, M. H., Diverging sex-morbidity trends in cancer of the mouth: hospital mordity study, *Br. Med. J.,* 2, 823, 1955.
7. Wood, C. A. P., The treatment of malignant disease of the face and jaws by radiation, *Br. Dent. J.,* 110, 234, 1961.
8. Phillips, A. J., Cancer mortality trends in Canada 1941-1958, *Br. J. Cancer,* 15, 1, 1961.
9. Anderson, D. L., Cause and prevention of lip cancer, *J. Can. Dent. Assoc.,* 37, 138, 1971.
10. Memorial Hospital for Cancer and Allied Diseases and the James Ewing Hospital of the City of New York: Statistical Report of End Results, 1949-1957.
11. Cooke, B. E. D., Recognition of oral cancer. Causes of delay, *Br. Dent. J.,* 142, 96, 1977.
12. Cowdry, E. V., *Cancer Cells,* W. B. Saunders, Philadelphia, 1955.
13. Bittner, J. J., Some possible effects of nursing on the mammary gland tumor incidence in mice, *Science,* 84, 162, 1936.
14. Shimkin, M. B., *Cancer: Diagnosis, Treatment and Prognosis,* 2nd ed., Ackerman, L. V. and del Regato, J. A., Eds., C. V. Mosby, St. Louis, 1954, chap. 2.
15. Pott, Sir Percival, *Chirurgical Observations Relative to the Cataract, the Polypus of the Nose, the Cancer of the Scrotum, the Different Kinds of Ruptures and the Mortification of the Toes and the Feet,* Hower, Clark and Pollins, London, 1775.
16. Wolff, G., *Chemical Induction of Cancer,* Harvard University Press, Cambridge, 1952.
17. Yamagiwa, K. and Ichikawa, K., Experimental study of the pathogenesis of carcinoma, *J. Cancer Res.,* 3, 1, 1918.
18. Kennaway, E. L. and Heiger, I., Carcinogenic substances and their fluorescence spectra, *Br. Med. J.,* 1, 1044, 1930.
19. Wynder, E. L., Bross, I. J., and Feldman, R. M., A study of the etiological factors in cancer of the mouth, *Cancer,* 10, 1300, 1957.
20. Wynder, E. L. and Stellman, S. D., Comparative epidemiology of tobacco related cancers, *Cancer Res.,* 37, 4608, 1977.
21. Graham, S., Dayal, H., Rohrer, T., Swanson, M., Sultz, H., Shedd, D., and Fischman, S., Dentition, diet, tobacco and alcohol in the epidemiology of oral cancer, *J. Natl. Cancer Inst.,* 59, 1611, 1977.
22. Jafarey, N. A., and Zaidi, S. H. M., Carcinoma of the oral cavity and oral pharynx in Karachi (Pakistan). An appraisal, *Trop. Doctor,* 6, 63, 1976.
23. Chapman, I. and Redish, C. H., Tobacco-induced epithelial proliferation in human subjects, *Arch. Pathol.,* 70, 133, 1960.
24. Hirayama, T., An epidemiological study of oral and pharyngeal cancer in Central and Southeast Asia, *Bull. W.H.O.,* 34, 41, 1966.
25. Salley, J., Experimental carcinogenesis in the cheek pouch of the Syrian hamster, *J. Dent. Res.,* 33, 253, 1954.
26. Dunham, L. J., Muir, C. S., and Hamner, J. E., Epithelial atypia in hamster cheek pouches treated repeatedly with calcium hydroxide, *Br. J. Cancer,* 20, 588, 1966.
27. Peacock, E. E., Greenberg, G. E., and Brawley, B. W., The effect of snuff and tobacco on the production of oral carcinoma: an experimental and epidemiological study, *Ann. Surg.,* 1, 542, 1960.
28. Kent, S., Spontaneous and induced malignant neoplasms in monkeys, *Ann. N.Y. Acad. Sci.,* 85, 819, 1960.
29. Cohen, B. and Smith, C. J., Aetiological factors in oral cancer: experimental investigation of early epithelial changes, *Helv. Odontol. Acta,* 11, 112, 1967.
30. Pindborg, J. J. and Renstrup, G., Studies in oral leukoplakias. II. Effect of snuff on oral epithelium, *Acta Derm. Venereol.,* 43, 271, 1963.
31. Zegarelli, E. V., Everett, F. G., and Kutscher, A. H., Familial white-folded dysplasia of the mucous membranes. An atlas of oral lesions, *Oral Surg. Oral Med. Oral Pathol.,* 14, 1436, 1961.
32. Hamner, J. E., Betel quid inducement of epithelial atypia in the buccal mucosa of baboons, *Cancer,* 30, 1001, 1972.
33. Hamner, J. E. and Reed, O. M., Betel quid carcinogenesis in the baboon, *J. Med. Primatol.,* 1, 75, 1972.

34. Hamner, J. E. and Reed, O. M., Betel quid inducement of carcinoma *in situ* in the buccal mucosa of baboons, *Med. Primatol.*, Proc. 3rd Conf. Exp. Med. Surg. Primates, Part II, Lyon, 1972, 169.

35. Hamner, J. E., Oral implantology and oral carcinogenesis in primates, *Am. J. Phys. Anthropol.*, 38, 301, 1973.

36. Di-Paulo, J. A. and Levin, M. J., Tumor incidence in mice after oral painting with cigarette smoke condensate, *J. Natl. Cancer Inst.*, 34, 595, 1965.

37. Saigopal, G., Narasimhan, B., Raju, M. V., and Reddy, C. R., Effect of tobacco tar on the palates of rats and mice, *Clinician*, 37, 370, 1973.

38. Bastiaan, R. J. and Reade, P. C., The histopathologic features which follow repeated applications of tobacco tar to rat lip mucosa, *Oral Surg. Oral Med. Oral Pathol.*, 49, 435, 1980.

39. Reddy, D. G. and Anguli, B. C., Experimental production of cancer with betel nut, tobacco and slake lime mixture, *J. Indian Med. Assoc.*, 49, 315, 1967.

40. Lacassagne, A. and Latarjet, R., Action of methylcholanthrene on certain scars on the skin of mice, *Cancer Res.*, 6, 183, 1946.

41. Levy, B. M., Gorlin, R., and Gottsegin, R., Histology of early reactions of skin and mucous membrane of the lip of the mouse to a single application of a carcinogen, *J. Natl. Cancer Inst.*, 12, 275, 1951.

42. Salley, J. J., Penetration of carcinogenic hydrocarbons into oral tissues as observed by fluorescence microscopy, *J. Dent. Res.*, 40, 177, 1961.

43. Wallenius, K., Effect of 7,12-dimethylbenzanthracene on epidermis and oral mucosa in the white rat, *Odontol. Revy*, 16, 204, 1965.

44. Wallenius, K., Experimental oral cancer in the rat with special reference to the influence of saliva, *Acta Pathol. Microbiol. Scand.*, 66 (Suppl. 180), 5, 1966.

45. Fujino, H., Chino, T., and Imai, T., Experimental production of labial and lingual carcinoma by local application of 4-nitroquinoline N-oxide, *J. Natl. Cancer Inst.*, 35, 907, 1965.

46. Lekholm, U. and Wallenius K., Experimental oral cancer in rats with xerostomia, *Odontol. Revy*, 27, 11, 1976.

47. Wallenius, K., Mathur, A., and Abdulla, M., Effect of different levels of dietary zinc on development of chemically induced oral cancers in rats, *Int. J. Oral Surg.*, 8, 56, 1979.

48. Levy, B. M., The experimental production of carcinoma of the tongue in mice, *J. Dent. Res.*, 37, 950, 1958.

49. Mesrobian, A. and Shklar, G., Gingival carcinogenesis in the hamster using tissue adhesives for carcinogen fixative, *J. Periodontol.*, 40, 603, 1969.

50. Carter, R. L., Heathcote, N., and Roe, F. J. C., Unusual susceptibility of lingual epithelium to tumour induction by 4-nitroquinoline-1-oxide, *Nature (London)*, 215, 549, 1967.

51. Fujita, K., Kaku, T., Sasaki, M., and Onoe, T., Experimental production of lingual carcinomas in hamsters by local application of 9,10-dimethyl-1,2-benzanthraciene, *J. Dent. Res.*, 52, 327, 1973.

52. Fujita, K., Kaku, T., Sasaki, M., and Onoe, T., Experimental production of lingual carcinomas in hamsters: tumor characteristics and site of formation, *J. Dent. Res.*, 52, 1176, 1973.

53. Marefat, P. and Shklar, G., Experimental production of lingual leukoplakia and carcinoma, *Oral Surg. Oral Med. Oral Pathol.*, 44, 578, 1977.

54. Marefat, P. and Shklar, G., Lingual carcinogenesis in an inbred strain of hamsters, *J. Oral Pathol.*, 7, 38, 1978.

55. Marefat, P. and Shklar, G., Lingual leukoplakia and carcinoma, an experimental model, *Progr. Exp. Tumor Res.*, 24, 259, 1979.

56. Giunta, J. and Shklar, G., Studies on tongue carcinogenesis in rats using DMBA with and without cyanoacrylate adhesive, *Arch. Oral Biol.*, 17, 617, 1972.

57. Lijinsky, W. and Taylor, H. W., Carcinogenicity of N-nitroso-3,4-dichloro- and N-nitroso-3,4-dibromopiperidine in rats, *Cancer Res.*, 35, 3209, 1975.

58. Lurie, A. G. and Cutler, L. S., Effects of low-level X-radiation on 7,12-dimethylbenz[a]anthracene-induced lingual tumors in Syrian golden hamsters, *J. Natl. Cancer Inst.*, 63, 147, 1979.

59. Shklar, G., Marefat, P., Kornhouser, A., Trickler, D. P., and Wallace, K. D., Retinoid inhibition of lingual carcinogenesis, *Oral Surg. Oral Med. Oral Pathol.*, 49, 325, 1980.

60. Levy, B. M. and Ring, J. R., Experimental production of jaw tumors in hamsters, *Oral Surg. Oral Med. Oral Pathol.*, 32, 262, 1950.

61. Green, G. W., Collins, D. A., and Bernier, J. L., Response of embryonal odontogenic epithelium in the lower incisor of the mouse to 3-methyl-cholanthrene, *Arch. Oral Biol.*, 1, 325, 1960.

62. Zegarelli, E. V., Adamantoblastomas in the Slye strain of mice. *Am. J. Pathol.*, 20, 23, 1944.

63. Mesrobian, A. Z. and Shklar, G., Experimental oral malignant lymphoma using alveolar socket carcinogen implantation, *J. Periodontol.*, 42, 105, 1971.

64. Sela, J. and Harary, D., Fibrosacroma induced by administration of methylcholanthrene into post-extraction sockets of mandibular molars in rats, *Experientia*, 35, 110, 1979.

65. **Herrold, K. McD.**, Odontogenic tumors and epidermoid carcinomas of the oral cavity. An experimental study in Syrian hamsters, *Oral Surg. Oral Med. Oral Pathol.*, 25, 262, 1968.

66. **Jasmin, J. R., Brocheriou, C., Matar, A., Morin, M., Jasmin, C., and Lafuma, J.**, Orofacial tumors induced in rats with radioactive cerium chloride, *Biomedicine*, 30, 265, 1979.

67. **Rous, P.**, Transmission of a malignant new growth by means of a cell-free filtrate, *JAMA*, 56, 198, 1911.

68. **Shope, R. E.**, A filterable virus causing a tumor-like condition in rabbits and its relationship to virus myxomatosum, *J. Exp. Med.* 56, 803, 1932.

69. **Kilham, L.**, Transformation of fibroma into myxoma virus in tissue culture, *Proc. Soc. Exp. Biol. Med.*, 95, 59, 1957.

70. **Kilham, L.**, Metastasizing viral fibromas of gray squirrels: pathogenesis and mosquito transmission, *Am. J. Hyg.*, 61, 55, 1955.

71. **Shope, R. E.**, Infectious papillomatosis of rabbits. With a note on the histopathology by E. Weston Hurst, *J. Exp. Med.*, 58, 607, 1933.

72. **Parsons, R. J. and Kidd, J. G.**, A virus causing oral papillomatosis in rabbits, *Proc. Soc. Exp. Biol. Med.*, 35, 441, 1936.

73. **DeMonbrun, W. A. and Goodpasture, E. W.**, Infectious oral papillomatosis of dogs, *Am. J. Pathol.*, 8, 43, 1932.

74. **Lucke, B.**, Carcinoma in the leopard frog. Its probable causation by a virus, *J. Exp. Med.*, 68, 457, 1938.

75. **Gross, L.**, A filterable agent, recovered from Ak leukemic extracts, causing salivary gland carcinomas in C_3H mice, *Proc. Soc. Exp. Biol. Med.*, 83, 414, 1953.

76. **Stanley, H. R., Dawe, C. J., and Law, L. W.**, Oral tumors induced by polyoma virus in mice, *Oral Surg. Oral Med. Oral Pathol.*, 17, 547, 1964.

77. **Stanley, H. R., Bear, P. N., and Kilham, L.**, Oral tissue alterations in mice inoculated with the Rowe substrain of polyoma virus, *Periodontics*, 3, 178, 1965.

78. **Lucas, R. B.**, Odontogenic Tumours in Polyoma Virus-Infected Mice, Proc. 4th Intl. Acad. Oral Pathol., Johannesburg, South Africa, 1969, 120.

79. **Main, J. H. P. and Dawe, C. J.**, Tumor induction in transplanted tooth buds infected with polyoma virus, *J. Natl. Cancer Inst.*, 36, 1121, 1966.

80. **Levy, B. M., Taylor, A. C., Hampton, S., and Thoma, G. W.**, Tumors of the marmoset produced by Rous sarcoma virus, *Cancer Res.*, 29, 2237, 1969.

81. **Cady, D. and Catlin, D.**, Epidermoid carcinoma of the gum. A 20-year survey, *Cancer*, 23, 551, 1969.

82. **Srivastava, S. P. and Sharma, S. C.**, Gingival cancer, *Indian J. Cancer*, 5, 89, 1968.

83. **Zussman, W. V.**, Experimental gingival carcinoma — a preliminary report, *Arch. Oral Biol.*, 10, 535, 1965.

84. **Zussman, W. V.**, Infiltrating gingival carcinoma in rats ingesting N-2-fluorenylacetamide, *Cancer Res.*, 26, 544, 1966.

85. **Al-Ani, S. and Shklar, G.**, Effects of a chemical carcinogen applied to hamster gingiva, *J. Periodontol.*, 37, 36, 1966.

86. **Mesrobian, A. Z. and Shklar, G.**, Gingival carcinogenesis in the hamster using tissue adhesives for carcinogen fixation, *J. Periodontol. Periodontics*, 40, 45, 1969.

87. **Frazell, E. L.**, Clinical aspects of tumors of the major salivary glands, *Cancer*, 7, 637, 1954.

88. **Lucas, R. B.**, *Pathology of Tumours of the Oral Tissues*, 3rd ed., Churchill Livingstone, Edinburgh, 1976.

89. **Berg, J. W., Hutter, R. V. P., and Foote, F. W., Jr.**, The unique association between salivary gland cancer and breast cancer, *JAMA*, 204, 771, 1968.

90. **Brown, A. M. and Frankel A.**, Leukocyte mediated cytotoxicity of chemically induced rat salivary gland neoplasms, *J. Oral Pathol.*, 3, 239, 1974.

91. **Wilson, R. H., DeEds, F., and Cox, A. J., Jr.**, The toxicity and carcinogenic activity of 2-acetylaminofluorene, *Cancer Res.*, 1, 595, 1941.

92. **Heiman, J. and Meisel, D.**, Tumors of the salivary and parathyroid glands in rats fed with 2-acetylaminofluorene, *Cancer Res.*, 6, 617, 1946.

93. **Zussman, W. F.**, Salivary and sebaceous gland alterations in rat ingesting N-2-fluorenylacetamide, *Arch. Pathol.*, 80, 278, 1965.

94. **Reuber, M. D.**, Endometrial sarcomas of the uterus and carcinosarcomas of the submaxillary salivary gland in castrated AXC strain female rats receiving N,N-fluorenyldiacetamide and norethandrolone, *J. Natl. Cancer Inst.*, 25, 1141, 1960.

95. **Parkin, R. and Neale, S.**, The effect of isoprenaline on induction of tumors by methylnitrosourea in the salivary and mammary glands of female Wistar rats, *Br. J. Cancer*, 34, 437, 1976.

96. **Rush, H. P., Baumann, C. A., and Maison, G. L.**, Production of internal tumors with chemical carcinogens, *Arch. Pathol.*, 29, 8, 1940.

97. Franseen, C. C., Aub, J. C., and Simpson, C. L., Experimental tumors in lymph nodes and in endocrine and salivary glands, *Cancer Res.*, 1, 489, 1941.
98. Steiner, P. E., Comparative pathology of induced tumors of the salivary glands, *Arch. Pathol.*, 34, 613, 1942.
99. Bauer, W. H. and Byrne, J. J., Induced tumors of the parotid gland, *Cancer Res.*, 10, 755, 1950.
100. Bauer, W. H. and Grand, N. G., Interactions of salivary gland virus and experimentally produced carcinomas of mouse salivary gland, *Cancer Res.*, 14, 768, 1954.
101. Standish, S. M., Early histologic changes in induced tumors of the submaxillary salivary glands of the rat, *Am. J. Pathol.*, 33, 371, 1957.
102. Glucksman, A. and Cherry, C. P., The induction of adenomas by the irradiation of salivary glands of rats, *Radiat. Res.*, 17, 186, 1962.
103. Glucksman, A. and Cherry, C. P., Sex difference and carcinogenic dosage in the induction of neoplasms and salivary glands of rats, *Br. J. Cancer*, 25, 212, 1971.
104. Glucksman, A. and Cherry, C. P., The effect of sex and thyroid hormones on the induction of cancers in the salivary glands of rats, *Br. J. Cancer*, 20, 760, 1966.
105. Shafer, W. G. and Muhler, J. C., Effect of gonadectomy and sex hormones on the structure of the rat salivary glands, *J. Dent. Res.*, 32, 262, 1953.
106. Shafer, W. G. and Muhler, J. C., The effect of desiccated thyroid, propylthiouracil, testosterone and fluorine on the submaxillary glands of the rat, *J. Dent. Res.*, 35, 922, 1956.
107. Levi-Montalcini, R., Morphologic and metabolic defects of the nerve growth factor, *Arch. Biol.*, 76, 387, 1965.
108. Chaudhry, A. P., Liposky, R., and Jones, J., Dose-response of submandibular glands to carcinogen pellets in rats and hamsters, *J. Dent. Res.*, 45, 1548, 1966.
109. Schmutz, J. A. and Chaudhry, A. P., Incidence of induced tumors in the rat submandibular gland with different doses of 7,12-dimethylbenz-(α)-anthracene, *J. Dent. Res.*, 48, 1316, 1969.
110. Turbinger, S. and Shklar, G., Variations in experimental carcinogenesis of submandibular gland in three strains of rats, *Arch. Oral Biol.*, 14, 1065, 1969.
111. Wigley, C. B., Amos, J., and Brookes, P., Different tumors induced by benzo(a) pyrene and 7,8-dihydrodiol injected into adult mouse salivary gland, *Br. J. Cancer*, 37, 657, 1978.
112. Wigley, C. B. and Carbonell, A. W., The target cell in the chemical induction of carcinomas in mouse submandibular gland, *Eur. J. Cancer*, 12, 737, 1976.
113. Kim, S. K., Spencer, H. H., Weatherbee, L., and Nasjleti, C. E., Changes in secretory cells during early stages of experimental carcinogenesis in the rat submandibular gland, *Cancer Res.*, 34, 2172, 1974.
114. Okada, Y., Effects of isoproterinol on chemical carcinogenesis with DMBA in mouse salivary glands, *J. Oral Pathol.*, 8, 340, 1979.
115. Chaudhry, A. P. and Schmutz, J. A., Effect of prednisolone and thalidomide on induced submandibular gland tumors in hamsters, *Cancer Res.*, 26, 884, 1966.
116. Anbari, N., Shklar, G. and Caldo, E., The effect of systemically administered cortisone on salivary gland carcinogenesis in the rat, *J. Dent. Res.*, 44, 1056, 1965.
117. Turbiner, S., Shklar, G., and Cataldo, E., The effect of cold stress on chemical carcinogenesis of rat salivary glands, *Int. Assoc. for Dent. Res.*, Preprinted abstracts, 46th General Meeting, Abs. No. 311, American Dental Association, Chicago, 1968, p. 115.
118. Sheehan, R. and Shklar, G., The effect of cyclophosphamide on experimental salivary gland neoplasia, *Cancer Res.*, 32, 420, 1972.
119. Shklar, G. and Sonis, S. T., The effect of methotrexate on experimental salivary gland neoplasia in rats, *Arch. Oral Biol.*, 20, 787, 1975.
120. Pisanty, S., Eisenberg, E., and Shklar, G., The effect of levamisole on experimental carcinogenesis of rat submandibular gland, *Arch. Oral Biol.*, 23, 131, 1978.
121. Ciapparelli, L., Retief, D. H., and Fatti, L. P., The effect of zinc on 9,10-dimethyl-1,2-benzanthracene (DMBA) induced salivary gland tumors in the albino rat — a preliminary study, *S. Afr. J. Med. Sci.*, 37, 85, 1972.
122. Rowe, N. H., Grammer, F. C., Watson, F. R., and Nickerson, N. H., A study of environmental influence upon salivary gland neoplasia in rats, *Cancer*, 26, 436, 1970.
123. Englander, A. and Cataldo, E., Experimental carcinogenesis in duct-artery ligated rat submandibular gland, *J. Dent. Res.*, 55, 229, 1976.
124. Sela, J., Azachi, C., Levij, I. S., and Ulmansky, M., Fibrosarcoma or squamous cell carcinoma of rat parotid after instillation of DMBA into the duct, *J. Dent. Res.*, 53, 1498, 1974.
125. Handler, A. H. and Shepro, D., Cheek pouch technology: uses and application, in *The Golden Hamster*, Iowa State University Press, 1968.
126. Shepro, D., Eidelhoch, L. P., and Patt, D., Lymph node responses to malignant homo- and heterografts in the hamster, *Anat. Rec.*, 136, 193, 1960.

127. Ferguson, J. W. and Smillie, A. C., Vascularization of premalignant lesions in carcinogen-treated hamster cheek pouch, *J. Natl. Cancer Inst.*, 63, 1383, 1979.
128. Woods, D. A. and Smith, C. J., Ultrastructure and development of epithelial cell pseudopodia in chemically induced premalignant lesions of the hamster cheek pouch, *Exp. Mol. Pathol.*, 10, 160, 1970.
129. Homburger, F., Chemical carcinogenesis in the Syrian golden hamster. A review, *Cancer*, 23, 313, 1969.
130. Homburger, F., Chemical carcinogenesis in Syrian hamsters, *Progr. Exp. Tumor Res.* 16, 152, 1972.
131. Shklar, G., Recent advances in experimental oral and salivary gland tumors, *J. Oral Surg.*, 28, 495, 1970.
132. Shklar, G., Experimental oral pathology in the Syrian hamster, *Progr. Exp. Tumor Res.*, 16, 518, 1972.
133. Salley, J. J., Experimental carcinogenesis in the cheek pouch of the Syrian hamster, *J. Den. Res.*, 33, 253, 1954.
134. Salley, J. J., Histologic changes in the hamster cheek pouch during early hydrocarbon carcinogenesis, *J. Dent. Res.*, 36, 48, 1957.
135. Morris, A. L., Factors influencing experimental carcinogenesis in the hamster cheek pouch, *J. Dent. Res.*, 40, 3, 1961.
136. Renstrup, G., Smulow, J., and Glickman, I., Effect of chronic mechanical irritation on chemically induced carcinogenesis in the hamster cheek pouch, *J. Am. Dent. Assoc.*, 64, 770, 1962.
137. Silberman, S. and Shklar, G., The effect of a carcinogen (DMBA) applied to the hamsters buccal pouch in combination with croton oil, *Oral Surg. Oral Med. Oral Pathol.*, 16, 1344, 1963.
138. Homburger, F., Mechanical irritation, polycyclic hydrocarbons and snuff: effects on facial skin, cheek pouch and oral mucosa in Syrian hamsters, *Arch. Pathol.*, 91, 411, 1971.
139. Cavalaris, C. J., Acomb, A., and Velme, M., The effect of vitamin A on DMBA-induced tumorigenesis in the epithelium of the golden Syrian hamster cheek pouch, *Pharmacol. Ther. Dent.*, 1, 96, 1971.
140. Dresser, D. W., Adjuvanticity of vitamin A, *Nature (London)*, 217, 527, 1968.
141. Levij, I. S. and Polliack, A., Lymphoma-like lesions induced in the hamster cheek pouch with topical vitamin A palmitate, *Pathol. Microbiol.*, 34, 282, 1969.
142. Shklar, G., Schwartz, J., Graw, D., Trickler, D. P., and Williams, K. D., Inhibition of hamster buccal pouch carcinogenesis by 13-*cis*-retinoic acid, *Oral Surg. Oral Med. Oral Pathol.*, 50, 45, 1980.
143. Rubin, D. and Levij, I. S., Suppression by vitamins D_2 and D_3 of hamster cheek pouch carcinoma induced with 9,10-dimethyl-1,2-benzathracene, *Pathol. Microbiol.*, 39, 446, 1973.
144. Eisenberg, E., Neoplasia following cessation of DMBA application to hamster buccal pouch, *J. Dent. Res.*, 56, 1430, 1977.
145. Shklar, G., The effect of cortisone on the induction and development of hamster buccal pouch carcinomas, *Oral Surg. Oral Med. Oral Pathol.*, 23, 241, 1967.
146. Giunta, J. and Shklar, G., The effect of antilymphocyte serum on experimental hamster buccal pouch carcinogenesis, *Oral Surg. Oral Med. Oral Pathol.*, 31, 344, 1971.
147. Polliack, A. and Levij, I. S., 9,10 dimethyl-1,2-benzanthracene carcinogenesis in the hamster cheek pouch. *Arch. Pathol.*, 90, 494, 1970.
148. Gillman, F., Hathorn, M., and Penn, J., Action of cortisone in cutaneous and pulmonary neoplasms induced in mice by cutaneous application of methylcholanthrene, *Br. J. Cancer*, 10, 394, 1956.
149. Engelbreth-Holm, J. and Asboe-Hansen, G., Effect of cortisone on skin carcinogenesis in mice, *Acta Pathol. Microbiol. Scand.*, 32, 560, 1953.
150. Baserga, R. and Shubik, P., The action of cortisone on transplanted and induced tumors in mice, *Cancer Res.*, 14, 12, 1954.
151. Boutwell, R. K. and Rusch, H. P., The effect of cortisone on the development of tumors, *Proc. Am. Assoc. Cancer Res.*, 1, 5, 1953.
152. Sabes, W. R., Chaudry, A. P., and Gorlin, R. J., Effects of cortisone on chemical carcinogenesis in hamster pouch and submandibular salivary gland, *J. Dent. Res.*, 42, 1118, 1963.
153. Levij, I. S., Durst, A., and Polliack, A., The effect of castration on chemical carcinogenesis in the cheek pouch of the male Syrian golden hamster, *Oral Surg. Oral. Med Oral Pathol.*, 28, 709, 1969.
154. Polliack, A., Charuzy, I., and Levij, I. S., The effect of oestrogen on 9,10-dimethyl-1,2-benzanthracene (DMBA)-induced cheek pouch carcinoma in castrated and non-castrated male Syrian golden hamsters, *Br. J. Cancer*, 23, 781, 1969.
155. Polliack, A., Charuzy, I., and Levij, I. S., The effect of testosterone on chemical carcinogenesis in the buccal pouches of castrated and intact male hamsters, *Pathol. Microbiol.*, 35, 348, 1970.
156. Shklar, G., Cataldo, E., and Fitzgerald, A. L., The effect of methotrexate on chemical carcinogenesis of hamster buccal pouch, *Cancer Res.*, 26, 2218, 1966.
157. Sheehan, R., Shklar, G., and Tennenbaum, R., Azathioprine effect on experimental buccal pouch tumors, *Arch. Pathol.*, 21, 264, 1971.

158. **Turbiner, S. and Shklar, G.,** Effect of fluorouracil on carcinogenesis of rat submandibular gland, *J. Dent. Res.,* 50, 987, 1971.

159. **Weathers, D. R. and Halstead, C. L.,** Histologic study of the effect of 5-fluorouracil on chemically induced early dysplasia of the hamster cheek pouch, *J. Dent. Res.,* 48, 157, 1969.

160. **Suri, K., Goldman, H. M., and Wells, H.,** Carcinogenic effect of a dimethyl sulphoxide extract of betel nut on the mucosa of the hamster buccal cheek pouch, *Nature (London),* 230, 383, 1971.

161. **Dunham, L. J. and Herrold, K. M.,** Failure to produce tumors in the hamster cheek pouch by exposure to ingredients of betel quid; histopathologic changes in the pouch and other organs by exposure to known carcinogens, *J. Natl. Cancer Inst.,* 29, 1047, 1962.

162. **Kendrick, F. J.,** Some effects of a chemical carcinogen and of cigarette smoke condensate upon hamster cheek pouch mucosa, *Health Sci.,* 24, 3698, 1964.

163. **Moore, C. and Miller, A. J.,** Effect of cigarette smoke tar on hamster cheek pouch, *Arch. Surg., (Chicago),* 76, 786, 1958.

164. **Tabah, E. J., Gorechi, Z., and Ritchie, A. C.,** Effects of saturated solutions of tobacco tars and of 9,10-dimethyl-1,2 benzathracene on the hamster's cheek pouch, *Proc. Am. Assoc. Cancer Res.,* 2, 254, 1957.

165. **Peacock, E. E., Jr., Greenberg, B. G., and Brawley, B. W.,** The effect of snuff and tobacco on the production of oral carcinoma. An experimental and epidemiological study, *Ann. Surg.,* 151, 542, 1960.

166. **Elzay, R. P.,** Effect of alcohol and cigarette smoke as promoting agents in hamster pouch carcinogenesis, *J. Dent. Res.,* 48, 1200, 1969.

167. **Friedman, A. and Shklar, G.,** Alcohol and hamster buccal pouch carcinogenesis, *Oral Surg. Oral Med. Oral Pathol.,* 46, 794, 1978.

168. **Ranadive, K. J., Ranadive, S. N., Shivapurkar, N. M., and Gothoskar, S. V.,** Betel quid chewing and oral cancer: experimental studies on hamsters, *Int. J. Cancer,* 24, 835, 1979.

169. **Kiseleva, N. S., Milievskaja, I. L., and Coklin, A. V.,** Development of tumours in Syrian hamsters during prolonged experimental exposure to nas, *Bull. W.H.O.,* 54, 597, 1976.

170. **Dunham, L. J., Sheets, R. H., and Morton, J.,** Proliferative lesions in cheek pouch and esophagus of hamster treated with plants from Curacao, Netherland Antilles, *J. Natl. Cancer Inst.,* 53, 1259, 1974.

171. **Folkman, J., Merler, E., and Abernathy, C.,** Isolation of a tumor factor responsible for angiogenesis, *J. Exp. Med.,* 133, 275, 1971.

172. **Mohammad, A. R. and Mincer, H. H.,** Dinitrochlorobenzene contact hypersensitivity in the hamster cheek pouch, *J. Oral Pathol.,* 5, 169, 976.

173. **Mohammad, A. R.,** Immunologic manipulation of DMBA tumorigenesis in hamster cheek pouch by DNCB contact hypersensitivity, *J. Oral Pathol.,* 8, 147, 1979.

174. **Eisenberg, E. and Shklar, G.,** Levamisole and hamster pouch carcinogenesis, *Oral Surg. Oral Med. Oral Pathol.,* 43, 562, 1977.

175. **Cottone, J. A., Kafrawy, A. H., Mitchell, D., and Standish, S. M.,** The effect of levamisole on DMBA-induced carcinogenesis in the hamster cheek pouch, *J. Dent. Res.,* 58, 629, 1979.

176. **Shklar, G., Eisenberg, E., and Flynn, E.,** Immunoenhancing agents and experimental leukoplakia and carcinoma of the hamster buccal pouch, *Prog. Exp. Tumor Res.,* 24, 269, 1979.

177. **Marshack, M., Toto, P., and Kerman, R.,** Immunotherapy of chemically-induced tumors in the hamster cheek pouch with dinitrochlorobenzene, *J. Dent. Res.,* 57, 625, 1978.

178. **Elzay, R. and Regelson, W.,** Effect of pyran copolymer on DMBA experimental oral carcinogenesis in the golden Syrian hamster, *J. Dent. Res.,* 55, 1138, 1976.

INDEX

A

P